Lecture Notes in Mathematics

Edited by A. Dold, B. Eckmann and F. Takens

Subseries: Fondazione C.I.M.E., Firenze
Adviser: Roberto Conti

1429

S. Homer A. Nerode
R.A. Platek G.E. Sacks
A. Scedrov

Logic and Computer Science

Lectures given at the 1st Session of the Centro Internazionale Matematico Estivo (C.I.M.E.) held at Montecatini Terme, Italy, June 20–28, 1988

Editor: P. Odifreddi

Springer-Verlag
Berlin Heidelberg New York London
Paris Tokyo Hong Kong Barcelona

Authors

Steven Homer
Department of Computer Science and Mathematics
Boston University, Boston, MA 02215, USA

Anil Nerode
Mathematical Sciences Institute
Cornell University, Ithaca, NY 14853, USA

Richard A. Platek
Odyssey Research Associates
301A Harris B. Dates Drive, Ithaca, NY 14850-1313, USA

Gerald E. Sacks
Department of Mathematics
Harvard University, Cambridge, MA 02138, USA

Andre Scedrov
Department of Mathematics
University of Pennsylvania, Philadelphia, PA 19104, USA

Editor

Piergiorgio Odifreddi
Dipartimento di Informatica, Università
Corso Svizzera 185, 10149 Torino, Italy

Mathematics Subject Classification (1980): 03B40, 03B20, 03B70, 03D15

ISBN 3-540-52734-6 Springer-Verlag Berlin Heidelberg New York
ISBN 0-387-52734-6 Springer-Verlag New York Berlin Heidelberg

Printed in Germany

Printing and binding: Druckhaus Beltz, Hemsbach/Bergstr.
2146/3140-543210 – Printed on acid-free paper

Preface

The C.I.M.E. Meeting on *Logic and Computer Science* was held in June 1988 in Montecatini, Italy. It was attended by some one hundred people from all over Europe, and it consisted of five short courses on mainstream aspects of Applied Logic.

In particular, the following fields were touched: foundational aspects of both logical (Sacks) and functional (Scedrov) programming languages; constructive logic (Nerode); complexity theory (Hartmanis and Homer); and program verification (Platek).

The present volume collects the lecture notes for those classes (with only one exception). We hope that they will turn out to be useful both to the people who attended the meeting, and to those who did not, but share with all of us an interest in the foundational aspects of Computer Science and the applications of Logic.

On behalf of the organization, I would like to thank the speakers and the participants for making the meeting a successful one.

Piergiorgio Odifreddi

TABLE OF CONTENTS

The Isomorphism Conjecture and Its Generalizations

Steven Homer*
Departments of Computer Science and Mathematics
Boston University
Boston, MA 02215
USA

This paper focuses on a particular problem in complexity theory, the isomorphism conjecture, which has been central to a large body of recent research. The problem was originally posed by Len Berman and Juris Hartmanis in [3]. Part of their the motivation for this problem is a theorem of John Myhill's from classical recursion theory and much of the work on the conjecture involves the interplay between recursion theory and complexity theory. Mathematical logic plays a major role in the definitions of the concepts and in indicating possible methods of solution. This paper will first present some background, including the original conjecture and first results concerning it. Then several generalizations of the conjecture and recent work concerning these generalizations will be discussed. Finally relativizations of the conjecture will be briefly explored. Throughout the paper the interaction with recursion theory and the many open problems which arise will be stressed.

1. The Isomorphism Conjecture

We begin with the work of Berman and Hartmanis [3]. They undertook the study of the structure of the NP-complete sets. There are literally thousands of such sets and, due to their practical importance, their study is one of the central topics of complexity theory. In [3] the question asked was, how similar are all of these NP-complete problems and what structure do they have in common ? Given that these many problems come from extremely disparate and unrelated areas of computer science they reached the surprising conclusion that all of the known (at that time) NP-complete sets are very similar, in fact essentially the same. More precisely they proved that they are all isomorphic via polynomial-time isomorphisms (p-isomorphic). They conjectured that *all* NP-complete problems are p-isomorphic.

*This work was supported in part by NSA grant #MDA904-87-H-2003 and by NSF grant #MIP-8608137.

Given the background provided by classical recursion theory this conjecture seems perfectly reasonable. The well known isomorphism theorem of John Myhill [18] states that all many-one complete recursively enumerable are isomorphic via a recursive isomorphism. If Myhill's proof worked in this subrecursive setting the conjecture would follow. However, as we will see, the subrecursive case does not follow from the recursive but rather presents subtleties and complications which are unique to complexity theory. Research on the isomorphism conjecture provides a good illustration of the difficulties present in the subrecursive setting which simply never arise in recursion theory. Surprisingly this research has some similarity to research in set recursion in generalized recursion theory (see Slaman [19]). A precise study of the parallels between these two areas might be worth pursuing.

Now to some definitions and terminology. All sets (problems) will be subsets of $\{0,1\}^*$. This papers deals exclusively with polynomial-time reducibilities. However, we will be careful to distinguish which type of polynomial reducibility we are using at any time. Four types of reducibilities will be used. They are Turing, $\leq_T^p$, truth-table, $\leq_{tt}^p$, many-one, $\leq_m^p$, and one-one, $\leq_{1-1}^p$, polynomial reducibility. Even among these four we will have occasion to distinguish several different restrictions and extensions. For any of these reducibilities, $\leq$, and any set S, the $\leq$-degree of S = $\{A \mid A \leq S \text{ and } S \leq A\}$. A set C is $\leq$-hard for a complexity class Q if, B ε Q implies B $\leq$ C. C is $\leq$-complete for Q if C ε Q and it is $\leq$-hard for Q. Readers unfamiliar with these reducibilities and their elementary properties might consult the paper of Ladner, Lynch and Selman [14].

NP-complete sets are sets which are $\leq_m^p$-complete for NP. A polynomial-time isomorphism (p-isomorphism) between two problems A and B is a one-one, surjective polynomial-time computable and invertible function from $\{0,1\}^*$ to $\{0,1\}^*$ which reduces A to B. Using these definitions, the conjecture that all NP-complete sets are p-isomorphic can now be precisely understood. Note that the isomorphism conjecture implies that P $\neq$ NP.

The first results in support of the conjecture are contained in the Berman, Hartmanis paper. There they show that many of the well-known NP-complete sets are p-isomorphic. Their method is to give conditions which imply p-isomorphism. They then show that many common NP-complete problems satisfy these conditions. The formulation of the result given here is due to Mahaney and Young [17]. It is essentially equivalent to the original one. Proofs of these results are omitted here, but can be found in [17]. SAT is, as usual, the collection of satisfiable Boolean formulas.

Definition: A polynomial padding function for a set S is is a one-one, polynomial time computable and invertible function p such that for all x and y, $x \varepsilon S$ if and only if $p(x,y) \varepsilon S$.

Theorem 1: Two sets which are in the same polynomial many-one degree both of which have padding functions are p-isomorphic.

Corollary 1: An NP-complete set A is p-isomorphic to SAT if and only if A has a padding function.

Proof of corollary (sketch): The corollary follows from Theorem 1 once a padding function for SAT is constructed. To pad a formula x with a binary padding string y, allocate a new literal for each 0 bit of y and the negation of a new literal for each 1 bit of y. Now form p(x,y) by conjoining x with the new literals and negations of literals.

The above Theorem gives a condition for p-isomorphism in terms of the sets structural properties. Often, as in the Corollary, we are given a particular "canonical" complete set (like SAT) and want to determine if another complete set is p-isomorphic to it. A second, related and often useful condition implying isomorphism can be given in terms of the reducibility properties of the sets. Let $\leq^p_{1-1,invertible}$ denote a reduction witnessed by a 1-1 polynomial-time computable and polynomial-time invertible function.

Theorem 2: Let C be a set with a polynomial padding function and B be such that B $\leq^p_{1-1,invertible}$ C and C $\leq^p_{1-1,invertible}$ B. Then B and C are p-isomorphic.

Proof: Let p be a polynomial padding function for C, f witness the invertible reduction from C to B and g witness the invertible reduction from B to C. We define a padding function q for B by, q(z,y) = f(p(g(z),y)). Then z ε B iff g(z) ε C iff p(g(z),y) ε C iff f(p(g(z),y))=q(z,y) ε B. It is straightforward to check that q is polynomial-time invertible, so it is a padding function for B. The p-isomorphism follows by Theorem 1.

Using Theorem 2 we can prove that or NP-complete sets, 1-1, invertible completeness is enough to ensure p-isomorphism with SAT. A similar result holds for any other $\leq^p_m$-degree which contains a complete set (like SAT) which is $\leq^p_{1-1,invetible}$-complete and has a polynomial padding function.

Corollary 2: Let B be NP-complete. Then SAT is p-isomorphic to B iff SAT $\leq^p_{1-1,invertible}$ B.

Using Corollary 1, most known NP-complete sets can be shown to be p-isomorphic. One possibly exceptional class of NP-complete sets, not all known to have padding functions, are the p-creative sets defined by Joseph and Young in [8]. These sets are the polynomial-time analogs of the recursively creative sets. They have a number of interesting properties. Past this, little is known concerning the original conjecture.

2. The Conjecture for Other Subrecursive Classes

To better focus on the difficulties concerning a solution to the isomorphism conjecture and to examine the efficacy of the recursion-theoretic methods which are so successful in the r.e. case, we next consider two other well-studied complexity classes, deterministic exponential time (E = $DTIME(2^{linear})$) and nondeterministic exponential time (NE = $NTIME(2^{linear})$). What form does the isomorphism conjecture take here ? We consider sets in (deterministic or nondeterministic) exponential time which are complete for many-one polynomial-time reductions. Such sets form a polynomial many-one degree. The isomorphism question is then, as before, whether that degree *collapses*. That is, whether all sets in the degree are p-isomorphic.

Another way to view this degree is that it is the polynomial many-one degree of the canonical complete set, which for E is K = {(e,x,k) | the eth exponential time set accepts x in ≤ k steps, k a binary integer}. It is straightforward to show that K is actually $\leq^{p}_{l.i.,invertible}$-complete for E.

The isomorphism question for these and other subrecursive classes has in recent years come under study and some interesting results have been achieved. In many ways the non-deterministic classes are the more interesting as they intuitively seem more similar to NP, not being closed under complement. We begin, however, with deterministic exponential time as here the results are the cleanest.

In order to motivate our approach let us first briefly review the progenitor of all of these ideas, the result of John Myhill [18]. Myhill proved that all r.e. many-one complete sets are recursively isomorphic. This can be viewed as an effective version of the Cantor-Schroder-Bernstein Theorem from set theory. Its proof has two parts.

First it is shown that all many-one complete r.e. sets are one-one complete. This follows from the fact that all many-one complete r.e. sets have padding functions. Next it is proved that any two one-one complete r.e. sets are recursively isomorphic. The proof here is a "back and forth" argument where the recursive isomorphism is constructed value by value using the one-one reductions between the two sets. However, the search required to find the next value of the isomorphism may be very (more than polynomially) long.

The analogous proof for subrecursive complete sets fails in several places. While, as we will see, polynomial many-one completeness does imply polynomial one-one completeness, the proofs of these facts are quite different in this setting. Furthermore, the results here are not as strong as in the recursion-theoretic case. The one-one reductions are not known to be polynomially invertible, or even length increasing in general. If invertibility of the reductions (or equivalently the existence of invertible padding functions) were shown then p-isomorphism would follow from the results in Section 1.

One very nice result for deterministic exponential time is due to Len Berman [2]. He proved,

Theorem 3: A set A which is many-one polynomial-time complete for deterministic exponential time is one-one polynomial time complete as well. Furthermore, the one-one reductions to A can be made to be length increasing.

Proof (from Ganesan and Homer [5]): Let A be any arbitrary $\leq^{p}_{m}$-complete set in E and let K be the complete set for E defined above. Since K is 1-1, length-increasing complete, it is enough to show that $K \leq^{p}_{1-1,li} A$. ($\leq^{p}_{1-1,li}$ denotes a one-one, length-increasing polynomial-time reduction.)

Let $f_1, f_2, \ldots$ be an ennumeration of all polynomial time computable functions, such that $\lambda i,x.\ f_i(x)$ can be computed in time $2^{O(|i|+|x|)}$. We construct a set M in E such that the reduction, say, f_j from M to A is 1-1-li on {j} × N. In addition, the set M is constructed so that the function, g(x) = (j,x) will be a reduction from K to M. The required 1-1-li reduction from K to A is then $f(x) = f_j(g(x))$.

The following program describes the set M.

1. input (i,x)
2. if $|f_i(i,x)| \leq |(i,x)|$
3. then accept (i,x) iff f_i (i,x) $\notin$ A.
4. else if there exists y < x such that f_i (i,x) = f_i (i,y)
5. then accept (i,x) iff y $\notin$ K
6. else accept (i,x) iff $x \varepsilon K$.

Lemma : M is in E.

Proof: Let us compute the time required for M on input (i,x). Note that computing $f_i(i,x)$ takes time $2^{O(|i|+|x|)}$. Since, there are only $2^{O(|(i,x)|)}$ strings of the form (i,y) less than (i,x) all of them can be computed in time $2^{O(|(i,x)|)}$. Hence the condition on line 5 of the algorithm can be performed in $2^{|(i,x)|}$ steps. There are only three cases where the decision to accept (i,x) is made.

Case 1: $|f_i(i,x)| \leq |(i,x)|$. In this we accept (i,x) iff $f_i(i,x)$ is not in A. This can obviously be done in time $2^{|(i,x)|}$ as A is in E.

Case 2: The condition on line 4 holds. Since |y| < |(i,x)| membership of y in K can be decided in time $2^{|(i,x)|}$.

Case 3: M accepts (i,x) iff $x \varepsilon K$. This can be done in time $2^{|(i,x)|}$.

It is clear from the above cases that M is in E.

Lemma: If f_j is a reduction from M to A, then f_j is 1-1-li on {j} × N. Moreover, g(x) = (j,x) is a reduction from K to M.

Proof: If f_j is not length increasing, it is not a reduction from M to A because of line 3 of the construction. So, f_j has to be length increasing. Suppose, f_j is not 1-1. Let x_2 be the least element such that for some $x_1 < x_2$, $f_j(j,x_2) = f_j(j,x_1)$. By definition of M, $(j,x_1) \varepsilon M$ iff $x_1 \varepsilon K$,$(j,x_2) \varepsilon M$ iff $x_1 \notin K$. So, f_j can not be a reduction from M to A, a contradiction. Hence f_j is 1-1-li on {j} × N. Note that $(j,x) \varepsilon M$ iff $x \varepsilon K$ from the way M is defined. The elements of the form (j,x) will always fall in Case 3 of the algorithm. Hence, g(x) = (j,x) is a reduction from K to M.

Lemma: $K \leq^{P}_{1-1,li} A$.

Proof: Define $f(x) = f_j(g(x))$. Clearly g is 1-1-li. Since f_j is 1-1-li on the range of g, f is 1-1 and length increasing as well. f is also computable in polynomial time. It is easy to check that f is a reduction from K to A.

The next (and final) step toward showing that all exponential time complete sets are p-isomorphic would be to show that the reductions can be made to be polynomial-time invertible. That is, many-one completeness implies one-one invertible completeness. In fact, if this were the case, Theorem 2 would yield a solution to the isomorphism problem in this setting. Unfortunately, whether the reductions can be strengthened to be invertible remains an open problem.

Next, a related and somewhat more tractable problem is considered. We are unable to determine the structure of the many-one polynomial degree of the complete set for E. Is there a polynomial $\leq_m^p$-degree of a set in E whose p-isomorphism class we can determine ? In particular, can we find such a degree that collapses or doesn't collapse?

It is not difficult to construct an infinite exponential time set A which is polynomially immune (that is, contains no infinite polynomial subset). Clearly such an A is not p-isomorphic to $A \times N$. However, $A \leq_m^p A \times N$ via f(x)=(x,1) and $A \times N \leq_m^p A$ via f(x,y)=x. So $A \times N$ and A are in the same $\leq_m^p$-degree and we have an example of a polynomial many-one degree which does not collapse. The construction of such a set A is given in Kurtz, Mahaney and Royer [11]. In fact, they construct such a set which is complete with respect to polynomial 2-truth-table reducibility, a slightly weaker reducibility than $\leq_m^p$. Intuitively A is 2-truth-table reducible to B ($A \leq_{2-tt}^p B$) if and only if there is a polynomial time computable function f such that for each x, f(x) is a list of at most two string w and z, and whether x is in A can be determined as a boolean functions of the answers to whether w and z are in B. Kurtz, Mahaney and Royer [11] show,

Theorem 4: There is a p-immune set in E which is 2-tt complete for E.

Until quite recently it was not known if any polynomial many-one degree (complete or not) was collapsing. The first such example is found in [11] where it is shown,

Theorem 5: There is a set A in E which is 2-tt complete for exponential time and whose polynomial many-one degree collapses.

The proof is not given here. It is a finite injury priority argument from recursion theory. Both of the above theorems hold for any deterministic subrecursive complexity classes containing E as well. There seems to be no way to strengthen the argument and make A many-one complete, hence settling the conjecture for exponential time.

For nondeterministic exponential time (NE) the situation is somewhat more difficult. This is due to the nondeterminism and in particular the fact that the class is probably not closed under complement. Nonetheless, this and other nondeterministic classes are in some respects the more interesting as it is, after all, a nondeterministic classes in which we are the most interested.

Only one part of the above theorems holds for nondeterministic classes as well. The next two results can be found in Ganesan and Homer [5].

Theorem 6: A set A which is many-one polynomial-time complete for nondeterministic exponential time is one-one polynomial time complete as well.

The proof of Theorem 6 is an extension of the methods used to prove Theorem 3 and is omitted. For NE the question of whether the reductions can be made length increasing is open and interesting. The best known result is the following. A function f is said to be *exponentially honest* if for all x, $2^{|f(x)|} \geq |x|$.

Theorem 7: The one-one reductions provided by Theorem 5 can be made to be exponentially honest.

The results of Kurtz, Mahaney and Royer for E do not apply in this case. No nondeterministic complete class for $\leq_{1i}^{p}$ or any weaker reducibility is known to collapse.

3. The Conjecture for the Polynomial Time Complete R. E. Sets

The setting of r.e. sets is an inviting one in which to work on the isomorphism conjecture. The original theorem concerning recursive isomorphisms was proved there and there are many methods and concepts which have been developed concerning r.e. sets and which might prove useful in this instance. Disappointingly, no more is known in this setting than for the NE sets. Still, it seems a fertile area for possible future progress on the isomorphism problem, one in which at least some of the problems associated with subrecursive classes disappear.

Recursively enumerable sets with respect to polynomial reducibilities were first examined in the manuscript of Dowd [4]. Here the isomorphism conjecture in this setting was first defined. It asks whether any two polynomial many-one complete for r.e. sets are polynomial-time isomorphic. One such complete set is the canonical r.e. complete set K = { (e,x) | the eth r.e. set accepts x} and the isomorphism problem here is the question; Are all sets in the polynomial many-one degree of K p-isomorphic ? Hence we are working here with well understood r.e. sets, but with respect to the same polynomial reducibilities. What's known is meager and looks all too familiar.

Theorem 8: A set A which is many-one polynomial-time complete for recursively enumerable sets is one-one polynomial time complete as well. Furthermore, the one-one reductions can be made to be exponentially honest.

The proof is omitted here. It is quite similar to the proofs of Theorems 3 and 6 and can be found in Ganesan and Homer [5].

Similarly the work of Kurtz, Mahaney and Royer [11] applies here in a manner somewhat weaker than the E case. It yields,

Theorem 9: There is a set A recursive in the halting problem which is 2-tt hard for r.e. sets and whose polynomial many-one degree collapses.

Here D being 2-tt-hard for r.e. means that for any r.e. set A, A $\leq^{P}_{2\text{-}tt}$ D, but not necessarily that D is r.e. Similarly, there is also a 2-tt complete for r.e. polynomial many-one degree which does not collapse. Its construction is similar to the subrecursive case construction mentioned above.

Finally, we note that the difficulty here is a real one. The next result of Dowd [4] says that settling the isomorphism question negatively for r.e. sets would answer the P =? NP question as well.

Theorem 10: If there are two non-p-isomorphic many-one complete for r.e. sets then $P \neq NP$.

Proof : Assume P=NP and let A be many-one complete for r.e. Since A $\leq^{P}_{1,invertible}$ K, it is sufficient to show K $\leq^{P}_{1,invertible}$ A. Let K $\leq^{P}_{1}$ A via h (from Theorem 7) and let g(x,y) be an invertible padding function for K. Now given x, compute a function f(x) as follows. Find the least z such that |h(g(x,z))| > |x|, call it z_0. z_0 exists since h and g are one-one. Let f(x) = h(g(x,z_0)). Then f is one-one, length-increasing and x ε K iff g(x,z_0)ε K iff f(x) ε A. So f reduces K to A. The P=NP assumption implies that f is polynomial-time computable. Finally, f has a polynomial-time inverse as, to find f^{-1}(w), binary search for the least v, |v| ≤ |w|, such that h(v)=w. This can be done in NP and hence in P by the assumption. If there is no such v then w ∉ (range(f)). If such a v is found, invert g on v to find x and z such that g(x,z)=v, and check if z is least such that |h(g(x,z))| > |x|. If such exist output x = f^{-1}(w), if not w ∉ (range(f)).

4. The Relativized Conjecture

Relativized computations are now briefly considered. By this we mean computations which are allowed to ask questions of a fixed "oracle" set during their operation. The question now asked is whether an oracle set can be constructed relative to which the isomorphism conjecture is true and similarly whether a set can be found relative to which the conjecture is false. Motivation for these questions comes not because they directly imply an answer to the conjecture or even necessarily tell us its likely outcome. Rather, they indicate which methods will not work and imply the likely difficulty of the unrelativized conjecture.

Most theorems and their proofs relativize directly to any oracle set. If, for example, an oracle is found relative to which the isomorphism conjecture is false (true), then any proof which relativizes to an arbitrary oracle set cannot be used to prove the conjecture true (false). Many open problems in complexity theory have been shown to have a relativization for which it is true and anothers for which it is false. This is generally regarded as evidence of their intractability, at least using presently known methods.

We now return to the original isomorphism conjecture for NP. The next theorem of Stuart Kurtz gives a relativization where the isomorphism conjecture is false. The proof can be found in [13].

Theorem 11: There is a recursive set C such that there exists two NP^C-complete sets which are not P^C-isomorphic.

It remains an interesting open problem to construct an oracle relative to which the isomorphism conjecture for NP-complete sets is true. The closest we can come is the following result of Homer and Selman [7]. $\Sigma_2^{P,D}$ denotes the second level of the polynomial time hierarchy relative to the oracle D.

Theorem 12: There is a recursive set D such that $P^D \neq NP^D$ and all $\Sigma_2^{P,D}$-complete sets are P^D-isomorphic.

Outline of Proof: The set D is constructed to have the following properties:
1. Relative to D every one-one polynomial time computable, length increasing function had a polynomial-time inverse,
2. $P^D \neq NP^D$, and
3. $\Sigma_2^{P,D} = \mathrm{DTIME}(2^{POLY})$.

1, 2, and 3 yield the Theorem as follows. Let A and B be two $\Sigma_2^{P,D}$-complete sets, so they are many-one equivalent. Theorem 4 above applies to DTIME(2^{POLY}) as well as E and implies that A and B are equivalent via one-one length increasing reductions. By 1, these one-one reductions are both polynomial-time invertible. Finally, the original Theorem of Berman and Hartmanis, suitably relativized, yields the p-isomorphism of A and B.

It seems somewhat strange that the above result comes so close to a positive relativization to the isomorphism conjecture yet cannot achieve it. Nonetheless it seems some other approach will be necessary to fully relativize the original conjecture. Some related results can be found in Goldsmith and Joseph [6].

I have steered a rather narrow course through some currently active areas surrounding the isomorphism conjecture and its generalizations. My path reflects the aspects of the research which have the most interaction with recursion theory and of course my own interests as well. I close with a few brief references to two other related, active research areas so that the reader may pursue them on their own.

The concept of one-way function has informally been mentioned several times already. For the purposes here, a one-way function is one which is easy (polynomial-time) to compute, length increasing and difficult (not polynomial-time) to invert. The implications between the existence of such functions and the isomorphism conjecture are strong and not yet completely understood. If there are no one-way functions, then all polynomial many-one complete sets for E are p-isomorphic. This follows from Theorem 3 and the argument outlined in Theorem 12. The converse of this statement is an interesting open problem. The situations the p-isomorphism question for NE or for NP vis-a-vis one-way functions are not as clearly understood. There are some partial results here. Some of them can be found in Joseph and Young [8] or in Watenabe [20].

The various problems considered here have all asked if certain polynomial many-one complete sets are p-isomorphic. Some progress has been made if more restrictive reducibilities are considered. That is, by considering more restrictive reducibilities there are less complete sets in a polynomial degree and hence more hope of proving them all to be p-isomorphic. A good example of this approach can be found in Allender [1]. There it is shown, for example, that all complete sets for PSPACE under one-way logspace reductions are p-isomorphic. Further examples of this approach can be found in [1] or in Ganesan and Homer [5].

References

1. Allender, E.W., Isomorphisms and 1-1 reductions, *Proceeding of the Structure in Complexity Theory Conference, Springer-Verlag Lecture Notes in Computer Science* (1986), pp. 12-22.

2. Berman, L., Polynomial reducibilities and complete sets, *Ph.D. Thesis, Cornell University,* 1977.

3. Berman, L. and J. Hartmanis, On isomorphism and density of NP and other complete sets, *SIAM J. Computing 6* (1977). pp. 305-322.

4. Dowd, M, On isomorphism, *manuscript,* 1978.

5. Ganesan, K. and S. Homer, Complete problems and strong polynomial reducibilities, Boston University Tech. Report #88-001, January 1988.

6. Goldsmith, J., and D. Joseph, Three results of the polynomial isomorphism of complete sets, *Proc. 27th Ann. IEEE Symp. Found. of Comp. Sci.,* 1896, 390-397.

7. Homer, S., and A. Selman, Oracles for structural properties: The isomorphism conjecture and public-key cryptography, *manuscript,* 1988.

8. Joseph, D., and P. Young, Some remarks on witness functions for non-polynomial and non-complete sets in NP, *Theor. Comp. Sci.,* 39, 1985, 225-237.

9. Ko, K., T. Long and D. Du, A note on one-way functions and polynomial-time isomorphisms, *Theoretical Computer Science,* 1986, to appear.

10. Ko, K. and D. Moore, Completeness, approximation and density, *SIAM J. Comp.* (1981), pp. 787-796.

11 Kurtz, S., S. Mahaney, and J. Royer, Collapsing Degrees, *Technical Report TR 86-008, University of Chicago,* October 1986.

12. Kurtz, S., S. Mahaney, and J. Royer, On Collapsing Degrees in NP, PSPACE and Σ_2^p, *Technical Report, University of Chicago,* 1987.

13. Kurtz, S., A relativized failure of the Berman-Hartmanis conjecture, *manuscript*, 1983.

14. Ladner, R., N. Lynch and A. Selman, A comparison of polynomial-time reducibilities, *Theor. Comp. Sci. 15*, 1981, 181-200.

15. Mahaney, S., On the number of p-isomorphism classes of NP-complete sets, *Proc. 22nd Ann. IEEE symp. Found. of Comp. Sci.*, pp. 271-278, 1981.

16. Mahaney, S., Sparse complete sets for NP: Solution of a conjecture of Berman and Hartmanis, *JCSS*, (1982), pp. 130-143.

17. Mahaney, S. and P. Young, Reductions among polynomial isomorphism types, *Theoretical Computer Science*, (1985), pp. 207-224.

18 Myhill, J., Creative sets, *Zeitshrift fur Math. Logik und Grundlagen der Mathematik* (1955), pp. 97-108.

19. Slaman, T., Aspects of E-recursion, *Ph.D. Thesis, Harvard University*, 1981.

20. Watenabe, O., On one-one p-equivalence relations, *Theoretical Computer Science*, (1985), pp. 157-165.

SOME LECTURES ON INTUITIONISTIC LOGIC

Anil Nerode[1]
Mathematical Sciences Institute
Cornell University
Ithaca, New York 14853, U.S.A.

INTRODUCTION

Intutionistic predicate logic has become an important tool for computer science. It makes possible an acute mathematical analysis of functional languages such as ML (Girard's Curry–Howard isomorphism, see Girard's thesis [1972], Leivant [1983a], or Howard [1980]) as well as first and higher order logic programming languages, and program development systems such as Constable's NUPRL. Due to the computational content of its existential quantifiers it is the logic of choice for description of many areas of computer science. How should we teach intuitionistic logic to students in mathematics and computer science? Classical logic texts explain classical predicate logic first and then go on to recursion theory, set theory, and model theory. Where in a logic course does intuitionistic logic fit? Since classical logic is what students know informally already from their early mathematics training, and we should build upon what is already known, we think classical logic should come first. We also think intuitionistic systems should come next.

But how should this subject be exposited to mathematics and computer science students? Let us examine briefly the emphasis of existing standard expositions, and explain why we thought another exposition was called for. If we look at philosophical intuitionist logic texts such as Dummett [1977], there is an emphasis on understanding Brouwer's philosophy as well as an exposition of the formal systems. Certainly Heyting's development of intuitionistic logic was intended to formalize accepted reasoning principles for Brouwerian philosophy. But the usefulness of intuitionistic formalism for computer science can be based on interpretations entirely within classical mathematics using full classical logic, interpretations which do not coincide with either Brouwer's or Heyting's. For an example, the classical mathematician can replace Heyting's interpretation of his formal intuitionistic logic using "Brouwerian contructions" with an alternate interpretation substituting computer programs for constructions. This is Kleene's recursive realizability (Kleene, [1952]). Understanding Brouwerian philosophy plays no more role in computer science than understanding Frege's philosophy plays in mathematics. So we do not recommend introducing the subject via Brouwer's and Heyting's philosophy beyond giving them full credit for an emphasis on using constructive proofs and for the discovery of systems of logic suitable for describing such proofs. Current mathematically oriented intuitionist logic texts such as those of Fitting [1983], Van Dalen [1986], Lambek and Scott [1986] emphasize set–theoretic, algebraic, topological, or categorical interpretations using such tools as consistency properties, prime filters in distributive lattices, open sets, complete Heyting algebras, and toposes. These expositions achieve set–theoretic, algebraic, topological or categorical clarity and elegance, but at the cost of computational content. For this reason we have chosen the development outlined here as an introduction to the subject.

The exposition below represents the first part of our undergraduate course for students of mathematics and computer science at Cornell. In these notes there is an obvious emphasis on concreteness and computational content, and specifically on acquisition of computational proof skills. The Cornell course starts out with correctness and completeness of classical Beth–Smullyan tableaux. This classical case is omitted here. Those who do not know this subject in concrete form should read the estimable first fifty pages of Smullyan [1968]. The course continues with the present notes. These notes deal with Kripke semantics thought of as representing "states of knowledge" about classical models. Generalizing the motivation for the Beth–Smullyan tableaux in the classical case with which the course started, an intuitionistic proof by tableaux is introduced as a failed search for a Kripke model counterexample. The tableaux used refine those of Fitting [1983] and Hughes and Cresswell [1968] and are well suited to hand or machine computation. In experienced hands these tableaux lead easily to nice systems of resolution–unification theorem proving for intuitionistic logic as in recent papers of Fitting. This latter topic is not covered here. The exposition also owes a lot to Van Dalen's text [1983] and his review article [1986]. Some suggestive recursion–theoretic examples are used as informal motivation. The formal meaning of these examples for intuitionist systems cannot be fully understood without a treatment of Kleene recursive realizability interpretations of higher order intuitionistic logic. We do not give such a treatment here. The second part of our course, not reproduced here, deals with abstract realizability interpretations of intuitionistic systems. These are defined using typed and untyped combinatory structures as in Beeson [1985]. In this latter part is the explanation of the place of the recursion–theoretic examples. The overall objective of that part is to explain the proofs of Girard's and Martin–Löf's normalization theorems, their relation to Gentzen cut–elimination theorems, and why theorems of this kind are the mathematical basis for interpreters and compilers for functional languages such as ML and proof systems such as NUPRL. But that story will have to wait for another day.

I am greatly indebted to James Lipton, Sloan Foundation Dissertation Fellow, for editorial improvements. I thank James Lipton and Hutchinson Fellow Duminda Wijesekera for their version of the decidability of intuitionistic propositional logic via tableaux.

INTUITIONISTIC LOGIC

Frege published a formal system of classical predicate logic in 1879. He gave a set of logical rules of deduction sufficient for all the logical reasoning in mathematics from Euclid to the present day. This of course includes the law of the excluded middle. It is the subject studied in standard textbooks on classical logic.

In 1935 Brouwer's student Heyting published a formal system of intuitionist predicate logic based on exactly the same logical connectives ($\wedge$, $\vee$, $\rightarrow$, $\exists$, $\forall$, T, F) as classical predicate logic. This intuitionistic propositional logic was also developed independently by Kolmogorov and Glivenko. Heyting's system was based, in analogy to Frege's codification of the reasoning of classical mathematics, on codifying the logical rules actually used in the development of intuitionistic mathematics. This system, of course, neither contains nor implies the unrestricted law of the excluded middle.

Gödel in his 1930 thesis proved the following completeness theorem for classical predicate logic.

Every classical predicate logic statement true in all intended interpretations (equivalently, relational systems) has a classical predicate logic proof.

For classical predicate logic a precise definition of interpretation and of truth of a statement in an interpretation was given later, in 1935, by Alfred Tarski. He based his definition of interpretation on the notion of relational system, developed by Schröder in the 1890's, together with a definition of truth of a statement in a relational system by induction on the definition of statement. In one of Tarski's 1935 approaches the relational system is specified by giving names to each of the objects in its domain, and by specifying for each n–ary relation symbol R which atomic statements $R(c_1, \ldots, c_n)$ are true. All other atomic statements are declared false. Tarski gives an inductive definition of truth or falsity for arbitrary statements with the above as base step for atomic statements.

We anticipate the discussion of Kripke's frame interpretations for intuitionistic predicate logic by emphasizing that in the base step of the Tarski definition of true statement for classical predicate logic mentioned above, every atomic statement not initially declared true is permanently declared false. The Tarski definition of truth for classical logic statements excludes evolutionary increase of the apprehension of truth of atomic statements for relational systems within the mathematician's mind, assuming instead complete knowledge of truth or falsity of atomic statements from the outset. In Kripke's frame interpretations for intuitionistic logic quite the opposite is the case. Indeed, this allowance for evolutionary increase of knowledge is the principal new feature introduced by Kripke.

Building on the 1950's work of Beth, in the early 1960's Kripke ([1965]) perfected analogues for intuitionistic predicate logic of relational systems for classical predicate logic. He called the analogue of a relational system a frame. It is essential to bear in mind that the definition of frame is a definition within classical mathematics. Kripke defined for frames an analogue to classical truth, the notion of forcing, and proved the following completeness theorem.

Every statement of intuitionistic predicate logic forced in all frames has an intuitionistic

predicate logic proof.

One virtue of frame semantics is that it makes it possible to manipulate the intuitionistic formalism with confidence, using only classical mathematical ideas. It can also be used to explain some of the connections between intuitionist inference and computer science. This semantics captures only a small part of Brouwer's ideas, some of those connected with "states of knowledge". Because the theory of frames relies on classical logic, frames are not accepted by "purists" in the intuitionist tradition. For a theory of complete frame semantics formulated in an intuitionistic metatheory, see Veldman [1976], De Swart[1976].

STATES OF KNOWLEDGE

In Brouwer's work, statements are called (intuitionistically) true only after they have been established by the mathematician. It is simply not known which statements (including atomic statements) will eventually be established. In the current state of knowledge, some have been established, some have not. Those which have not been established yet are not asserted to be (intuitionistically) true. Thus what is established is dependent on the current "state of knowledge". The statement "$\pi > 3$" was established before three hundred B. C., but "π is transcendental" was established only in the nineteenth century. What is established depends on the course of development that occurred among the possible courses of development that might have occurred. In the nineteenth century, Hermite established that "e is transcendental" before he established "π is transcendental", but we can imagine a possible course of development in which establishment of transcendence of these two numbers was in the reverse order. The actual course of development is one of many possible courses of development. A "state of knowledge" may give rise to many later "states of knowledge", corresponding to possible future courses of development into later states of knowledge where more may have been established.

Numerical Computation. "States of knowledge" arise in numerical computations in which the estimates always improve when they are changed. Envisage a computation in which one or more real numbers or numerical functions are approximated more and more closely in the course of a computation. In a computation for π by a series, at an early stage of the computation (early state of knowledge), we may establish the fact that $3.1 < \pi < 3.2$. This and similar assertions of degree of approximation about other numbers and functions known from the computation at this stage can be represented as atomic statements. The collection of all such currently established atomic statements asserting degree of approximation for numbers and numerical functions can be regarded as a "state of knowledge". At a later stage this knowledge may be improved by the computation, and it may be established that $3.14 < \pi < 3.15$, giving rise to an additional atomic statement which is part of that "state of knowledge". Of course since we are not perverse we assume that we only improve, and never degrade, our previous estimates by adding more established atomic statements. Many computational procedures are interactive and allow user choices to guide the course of the computation at various points in the computation. So a given "state of knowledge" embodied by a set of such atomic propositions may represent the information explicitly available about numbers and numerical functions in a possible state of a computation. Each such possible state may lead eventually through a series of steps to a number of later possible states determined by the available range of user choices. If, contrary to assumption, we allow in our "state of knowledge" atomic propositions which are not known to be valid but are tentative and will be withdrawn at a later point, we are outside the domain of applicability of Kripke frames and intuitionistic logic.

Non–Erasable Databases. "States of knowledge" arise naturally for "write only" databases made up of atomic statements, as is often the case with the PROLOG databases, and as could be the case for the numerical computations of the previous paragraph. More and more atomic statements are added to the database as they become available, no atomic statements are ever removed. The "write only" requirement is not much of a limitation. If an argument of each atomic fact recorded in the database is an annotation of unique conditions of entry of that atomic fact into the database (say by time and source), then the "write only" requirement may be imposed without restricting updating and without removing any previous atomic statements in the database. This is an accurate description of many optical disks used for storage at this time (1989). The "state of knowledge" is the entire set of atomic propositions making up the database, a later "state of knowledge" is one with additional atomic propositions. Integrity and other constraints limit what states of knowledge are possible and which ones can eventually succeed a given state of knowledge through a series of intermediate stages. If we allow the database to have entries removed or changed, the frame model below does not apply, and the rules of deduction of intuitionistic logic are not an appropriate tool for the subject.

FRAME SEMANTICS

Before defining (Kripke) frames we make precise what is meant by a language.

DEFINITION. A **language** $\mathbf{L} = <C, R, F>$ is given by the following information. A set C of constant symbols $c_1, c_2, \ldots$, a set R of relation symbols $R_1, R_2, \ldots$ of different arities, and a set F of function symbols $f_1, f_2, \ldots$, also of different arities.

For example, $\mathbf{L} = <\{0, 1\}, \{\leq\}, \{+, \cdot\}>$, the language of arithmetic, consists of constant symbols 0, 1, a relation symbol $\leq$ of arity 2, and function symbols $+$ and $\cdot$, both of arity 2. Often we just write $L = < 0, 1, \leq, +, \cdot >$.

Unless otherwise stated, the **logical symbols** will be $\wedge$, $\vee$, $\neg$, $\rightarrow$, F , $\forall$, $\exists$. A predicate logic over L is the collection of terms and well–formed formulas built up from the language **L** and the logical symbols. In most of our discussion of frames, we will consider only languages without function symbols.

The notion of intuitionistic frame formalizes the concept of "states of knowledge".

DEFINITION. Let a language **L** without function symbols be given. A **frame** over the language L, or simply an **L–frame**, $\mathcal{F} = (P, \leq, C(p), A(p))$ is a partially ordered set $(P, \leq)$, together with an assignment to each p in P of a pair $(C(p), A(p))$ consisting of a non–empty set $C(p)$ of individual constants which includes names for all the constants in **L**, and a (possibly empty) set $A(p)$ of atomic statements based on these constants and the relation symbols in L such that $p \leq q$ implies $C(p) \subseteq C(q)$ and $A(p) \subseteq A(q)$. We assume that the 0–arity predicate letter (propositional letter) F, denoting falsity, is not in $A(p)$. (Formulation of frame semantics in an intuitionistic metatheory seems to require that this last assumption be dropped. See Veldman, de Swart, op. cit.)

REMARK. Often $p \leq q$ is read "q extends p", or "q is a future of p". The elements of P are called possible worlds, or states of knowledge.

We now define the forcing relation for frames.

DEFINITION OF FORCING FOR FRAMES. Let p be in P and let φ be a sentence of the language $C(p)$. We give a definition of "p forces φ" by induction on the definition of sentence φ.

1) For atomic statements S, p forces S if and only if S is in $A(p)$.

2) p forces $(A \rightarrow B)$ if and only if
for all $q \geq p$, q forces A implies q forces B.

3) p forces $\neg A$ if and only if
for all $q \geq p$, q does not force A.

4) p forces $(\forall x)P(x)$ if and only if
for all $q \geq p$ and for every constant c in $C(q)$, q forces $P(c)$.

5) p forces $(\exists x)P(x)$ if and only if
there is a constant c in $C(p)$ such that p forces $P(c)$.

6) p forces $(A \wedge B)$ if and only
p forces A and p forces B.

7) p forces $(A \vee B)$ if and only if
p forces A or p forces B.

DEFINITION. Let φ be a sentence over a language L. We say that φ is forced in the L–frame $\mathcal{F}$ if every p in P forces φ. We say φ is intuitionistically valid if it is forced in every L–frame.

REMARK. Let M(p) be the classical model with domain C(p) and with true atomic statements exactly those in A(p). As we go from p to a $q > p$, we think of going from a classical model M(p) associated with p to a larger classical model M(q) associated with q with more atomic statements classically true, and therefore less atomic statements classically false. Clauses 1), 5), 6), 7) for the base step, and, or, there exists, are exactly as in the definition of truth in M(p). Classical truth of φ in M(p) and p forces φ don't generally coincide. But they do in an important case.

DEGENERACY PROPOSITION. In case there is only one state of knowledge p, "φ is classically true in M(p)" coincides with "p forces φ".

PROOF. The clauses in the definition of "p forces φ" coincide in this case with those for "φ is true in M(p)". In clauses 2, 3, and 4, the dependence on future states of knowledge is vacuous. For example, consider $A \rightarrow B$ iff ($\forall q \geq p$, q forces A implies q forces B). Since $q \geq p$ says q is p, this reduces to (p forces $A \rightarrow B$ iff p forces A implies p forces B).

REMARK. Clauses 2), 3), 4) have a quantifier ranging over elements of the partial ordering, namely "for all q, if $q \geq p$, then...". Clause 2) says p forces an implication only if in any greater state of knowledge q, if q forces the antecedent φ, then q forces the consequent ψ. This is a sort of permanence for implication in the face of more knowledge. Clause 4) says p forces the negation of φ when no greater state of knowledge forces φ. This says that φ cannot be forced by supplying more knowledge than p supplies. Since F is never forced (it occurs nowhere in the definition of forcing),

LEMMA. p forces $\varphi \rightarrow F$ if and only if p forces $\neg\varphi$.

Clause 4) says p forces a universally quantified statement implies that in all greater states of knowledge all instances of the statement are forced. This is a permanence of forcing universal statements in the face of any new knowledge beyond that supplied by p.

REMARK. The usual notation for "p forces A" is "$p \Vdash A$". This comes from Paul Cohen's work in set theory on the independence of the axiom of choice and the continuum hypothesis, not from intuitionistic tradition. Many people say "A is true at p" instead of "p forces A", but this invites confusion with classical truth in M(p).

REMARK. The definitions of $p \Vdash \varphi \rightarrow \psi$, $p \Vdash \neg\varphi$, $p \Vdash (\forall x)\varphi(x)$ call for every $p' \geq p$ to have a property. These quantifiers over P are classical predicate logic universal quantifiers over P in the same way that in classical logic $(\forall x)$ is a quantifier over a domain. They are simply not explicit in the statements.

RESTRICTION LEMMA. Let $\mathscr{F} = (P, \leq, C(p), A(p))$ be a frame, let p_0 be in P and let $P_{p_0} = \{q \in P \mid q \geq p_0\}$. Then

$$\mathscr{F}_{p_0} = (P_{p_0}, \leq, C(p), A(p))$$

is a frame, where $\leq$ and the function A and C are restricted to P_{p_0}. Also for q in P_{p_0}, q forces A in P iff q forces A in P_{p_0}.

This says the past does not count in forcing, only the future.

PROOF. By induction on the length of formulas.

EXERCISE. Prove the restriction lemma by induction on the definition of forcing.

Due to the degeneracy lemma, every classical model is (a one element partially ordered set) frame model in which forcing and classical truth coincide. A statement is classically valid if true in all classical models. Therefore

LEMMA. Any intuitionistically valid statement is classically valid.

It remains to see which classically valid statements are intuitionistically valid and which are not. We show how to verify by frame examples that some classically valid statements are not intuitionistically valid.

EXAMPLE 1. As expected, the statement $\varphi \vee \neg\varphi$ is not intuitionistically valid. Let the frame be

$$\begin{array}{c}\{\varphi\}\\ |\\ \{\,\}\end{array} \qquad \text{(which we will heretofore display as } \begin{array}{c}\varphi\\ |\end{array}\text{)}$$

Let 0 be the lower node, 00 be the upper node. Here the partially ordered set consists of two elements, the lower 0 with A(0) empty, the upper 00 with A(00) containing only φ. This is propositional logic so no domains C(p) are needed. Consider whether 0 forces $\varphi \vee \neg\varphi$. Certainly 0 does not force φ since it forces nothing. Certainly 0 does not force $\neg\varphi$ since 0 has the extension 00 forcing φ. So by definition 0 does not force $\varphi \vee \neg\varphi$ and this statement is not valid.

EXAMPLE 2. The statement $(\neg\varphi \to \neg\psi) \to (\psi \to \varphi)$ is not intuitionistically valid. Let the frame be

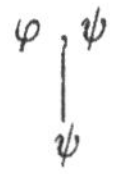

Let 0 be the lower node, 00 the upper node to get a partially ordered set. This is propositional logic, so no domains C(p) are needed. Suppose 0 forces $(\neg\varphi \to \neg\psi) \to (\psi \to \varphi)$. If so, then 0 forces $(\neg\varphi \to \neg\psi)$ would imply 0 forces $(\psi \to \varphi)$. Now 0 does force $(\neg\varphi \to \neg\psi)$ because neither 0 nor 00 force $\neg\varphi$. But 0 does not force $(\psi \to \varphi)$ because 0 forces ψ and does not force φ.

EXAMPLE 3. The statement $(\varphi \to \psi) \vee (\psi \to \varphi)$ is not intuitionistically valid. Let the partial ordering be

.

Let 0 be the lower node, let 00 be the upper left node, let 01 be the right upper node. Thus 0 forces nothing, 00 forces φ, 01 forces ψ. Since there is a node above 0, namely 00, which forces φ and not ψ, 0 does not force $\varphi \to \psi$. Similarly, 0 does not force $\psi \to \varphi$. So 0 does not force $(\varphi \to \psi) \vee (\psi \to \varphi)$.

EXAMPLE 4. The statement $\neg(\forall x)\varphi(x) \to (\exists x)\neg\varphi(x)$ is not intuitionistically valid. Let the partially ordered set be

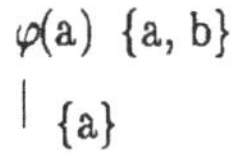

Since the statement is in predicate logic, domains are required.
Let 0 be the lower node, 00 the upper node. Here 00 forces the atomic statement $\varphi(a)$, C(0) is {a}, C(00) is {a, b}. Does 0 force $\neg(\forall x)\varphi(x) \to (\exists x)\neg\varphi(x)$? Now 0 does force $\neg(\forall x)\varphi(x)$ because neither 0 forces $\varphi(a)$ nor 00 forces $\varphi(b)$. But 0 does not force $(\exists x)\neg\varphi(x)$ because the only constant in C(0) is a and 0 does not force $\varphi(a)$. So $\neg(\forall x)\varphi(x) \to (\exists x)\neg\varphi(x)$ is not valid in frames.

EXAMPLE 5. Consider $(\forall x)(\varphi \vee \psi(x)) \to \varphi \vee (\forall x)\psi(x)$, x not free in φ. This is not intuitionistically valid.
Let the frame be

$\varphi, \psi(a)$, $\{a, b\}$

|

$\psi(a)$ $\{a\}$

Let 0 be the lower node, 00 be the upper node. Here 0 forces $\psi(a)$, C(0) is $\{a\}$, 00 forces φ, C(00) is $\{a, b\}$. Now 0 forces $(\forall x)(\varphi \vee \psi(x)) \rightarrow \varphi \vee (\forall x)\psi(x)$ would say that 0 forces $(\forall x)(\varphi \vee \psi(x))$ implies 0 forces $\varphi \vee (\forall x)\psi(x)$. But 0 does force $(\forall x)(\varphi \vee \psi(x))$ because 0 forces $\psi(a)$ and because 00 forces φ. However, 0 does not force $\varphi \vee (\forall x)\psi(x)$ because 0 does not force φ and 0 does not force $(\forall x)\psi(x)$ because 0 does not force $\psi(b)$.

REMARK. Let us summarize the use of pictures of partial orders above. We draw finite partial orderings as graphs whose nodes are the elements p of the partial order and with a branch drawn from p to q if q is an immediate successor to p, that is if $p \leq q$ and there is no r such that $p < r < q$. Smaller elements of the partial ordering are displayed below on the paper. Such a representation is called the Hasse diagram of the partial order. Drawing Hasse Diagrams is a very useful way of constructing partial orderings meeting given requirements. For frames each node p is labelled by the atomic statements in **A**(p) and the list of elements of C(p).

Let **P** be the partially ordered set of all finite sequences of non–negative integers with the partial ordering $\sigma \leq \tau$ if and only if σ is an initial segment of τ (So $001 \leq 0011$, but not $001 \leq 000$). In frame examples we use always partial orderings P which are initial segments of **P** – i.e., P contains with any element q any smaller element p. Equivalently P contains with any sequence q all of its initial segments p. These initial segments P of **P** are sufficient for the later completeness proof.

Here is the single most useful fact about forcing. It expresses the stability of forcing as one climbs up the partial ordering.

MONOTONICITY LEMMA. If p forces A and $q \in P$ and $q \geq p$, then q forces A.

PROOF. Let $\varphi(A)$ be the assertion that for all p, if p forces A and $q \geq p$ then q forces A. We show by induction on the logical complexity of A that for all formulas A, $\varphi(A)$. The inductive hypothesis is not used to verify the conclusion for clauses 2), 3), and 4). The clauses define the meaning of the connectives implication, negation, and universal quantification exactly to make this work. The induction hypothesis is used for the clauses 5), 6), 7) for or, and, there exists.

1) If A is atomic and p forces A, then A is in A(p). But $A(p) \subseteq A(q)$, so A is in A(q) and by definition q forces A.

2) Suppose p forces $A \to B$, and $q \geq p$. We show that q forces $A \to B$ by showing that if $r \geq q$, then r forces A implies r forces B. But transitivity says $r \geq p$, so that p forces $A \to B$ and r forces A implies that r forces B.

3) Suppose p forces $\neg A$ and $q \geq p$. We show q forces $\neg A$ by showing that if $r \geq q$ then r does not force A. But by transitivity, $r \geq p$, and this and the fact that p forces $\neg A$ imply r does not force A.

4) Suppose p forces $(\forall x)A(x)$ and $q \geq p$. We show q forces $(\forall x)A(x)$ by showing that for any $r \geq q$, we have for any c occurring in C(r) that r forces A(c). But by transitivity, $r \geq p$, so p forces $(\forall x)A(x)$ implies that for any c in C(r), r forces A(c).

5) Suppose p forces $A \vee B$, and $q \geq p$. Then by the definition of forcing either p forces A or p forces B. By the inductive hypothesis, namely that the theorem holds for A, B, we get that either q forces A or q forces B. This says by the definition of forcing that q forces $A \vee B$.

6) Suppose p forces $(A \wedge B)$, and $q \geq p$. Then by definition of forcing p forces A and p forces B. By the inductive hypothesis, q forces A and q forces B. Thus q forces $(A \wedge B)$.

7) Suppose p forces $(\exists x)A(x)$ and $q \geq p$. Then by the definition of forcing there is a c in C(p) such that p forces A(c). By the inductive hypothesis $q \geq p$ and p forces A(c) implies q forces A(c). Therefore q forces $(\exists x)A(x)$.

Monotonicity says the addition of new atomic statements at later states of knowledge q will not change forcing at earlier states of knowledge This monotone character distinguishes "truth" in an intuitionistic frame from "truth" in "non–monotonic logics", currently widely discussed in computer science. In these latter logics, statements forced at state of knowledge p are allowed to be unforced at states of knowledge $q > p$. In frames as time evolves, we remember all facts and only gain more knowledge.

DOUBLE NEGATION LEMMA. p forces $\neg\neg\varphi$ if and only if for any $q \geq p$ there is an $r \geq q$ such that r forces φ.

PROOF. p forces $\neg\neg\varphi$ if and only if every $q \geq p$ fails to force $\neg\varphi$, or if and only if every $q \geq p$ has an $r \geq q$ forcing φ.

We verify directly from definition the intuitionistic validity of some classically valid formulas. The monotonicity lemma is useful for this purpose.

EXAMPLE 6. $\varphi \to \neg\neg\varphi$ is intuitionistically valid. To see that any p forces $\varphi \to \neg\neg\varphi$ we assume that $q \geq p$ and q forces φ. We must show that q forces $\neg\neg\varphi$, or, by the double negation lemma, that for every $r \geq q$ there is an $s \geq r$ such that s forces φ. By transitivity, all such $s \geq q$, so by the monotonicity lemma s forces φ.

EXAMPLE 7. $\neg(\varphi \wedge \neg\varphi)$ is intuitionistically valid. To see that any p forces $\neg(\varphi \wedge \neg\varphi)$ we have to see that no $q \geq p$ forces $\varphi \wedge \neg\varphi$, or equivalently no $q \geq p$ forces both φ and $\neg\varphi$. For suppose that q forces both φ and $\neg\varphi$. Now q forces $\neg\varphi$ means no $r \geq q$ forces φ. Since $q \geq q$, we have both q forces φ and q does not force φ, a contradiction, and there is no such q.

EXAMPLE 8. $(\exists x)\neg\varphi(x) \to \neg(\forall x)\varphi(x)$ is intuitionistically valid. To see that any p forces $(\exists x)\neg\varphi(x) \to \neg(\forall x)\varphi(x)$, we need to show that if $q \geq p$ and q forces $(\exists x)\neg\varphi(x)$, then q forces $\neg(\forall x)\varphi(x)$. But q forces $(\exists x)\neg\varphi(x)$ says there is a c occurring in C(q) such that q forces $\neg\varphi(c)$. By monotonicity, any $r \geq q$ forces $\neg\varphi(c)$ too, so such an r does not force $(\forall x)\varphi(x)$, so q forces $\neg(\forall x)\varphi(x)$. Compare with example 4.

EXAMPLE 9. $\neg(\exists x)\varphi(x) \to (\forall x)\neg\varphi(x)$ is intuitionistically valid. To see that any p forces $\neg(\exists x)\varphi(x) \to (\forall x)\neg\varphi(x)$ we have to show that for any $q \geq p$, if q forces $\neg(\exists x)\varphi(x)$, then q forces $(\forall x)\neg\varphi(x)$. Now q forces $\neg(\exists x)\varphi(x)$ says that for every $r \geq q$, every c in C(r), r does not force $\varphi(c)$. By transitivity $s \geq r$ implies $s \geq q$. So for every $r \geq q$, every c occurring in C(r), no $s \geq r$ forces $\varphi(c)$. This says q forces $(\forall x)\neg\varphi(x)$.

EXAMPLE 10. If x is not free in φ, then $\varphi \vee (\forall x)\psi(x) \to (\forall x)(\varphi \vee \psi(x))$ is intuitionistically valid. To see that any p forces $\varphi \vee (\forall x)\psi(x) \to (\forall x)(\varphi \vee \psi(x))$ we must show that for any $q \geq p$, q forces φ or q forces $(\forall x)\psi(x)$ implies q forces $(\forall x)(\varphi \vee \psi(x))$. There are two cases. If q forces φ, then for any $r \geq q$, any c in C(r), q forces $\varphi \vee \psi(c)$, so q forces $(\forall x)(\varphi \vee \psi(x))$. If q forces $(\forall x)\psi(x)$, then for all $r \geq q$, all c occurring in C(r), r forces $\psi(c)$, so r forces $\varphi \vee \psi(c)$. This says q forces $(\forall x)(\varphi \vee \psi(x))$. Compare with example 5.

ADDITIONAL LEMMAS ON FORCING

WEAK QUANTIFIER LEMMA. 1) p forces $\neg(\exists x)\neg\varphi(x)$ if and only if for all $q \geq p$ and for all c in C(q) there is an $r \geq q$ such that r forces $\varphi(c)$. 2) p forces $\neg(\forall x)\neg\varphi(x)$ if and only if for all $q \geq p$, there exists an $s \geq q$ and a c in C(s) such that s forces $\varphi(c)$.

Proof of 1). This follows immediately from the definition.
Proof of 2). q forces $(\forall x)\neg\varphi(x)$ if and only if for all $r \geq q$ and all c in C(r) there is no $s \geq r$ such that s forces $\varphi(c)$. q does not force $(\forall x)\neg\varphi(x)$ if and only if there is an $r \geq q$ and a c in C(r) such that for some $s \geq r$, s forces $\varphi(c)$. So p forces $\neg(\forall x)\neg\varphi(x)$ if and only if for all $q \geq p$, there is an $r \geq q$ and a c in C(r) such that for some $s \geq r$, s forces $\varphi(c)$. But by

transitivity $s \geq q$ and c is also in C(s).

GENERALIZED DEGENERACY LEMMA. If p is a maximal element of P, then "A is true in M(p)" coincides with "p forces A". (Recall, M(p) is the classical model with domain C(p) and true atomic statements precisely those in A(p).)

EXERCISE SET 1.

Verify that the following classically valid statements are intuitionistically valid by direct argument with frames. Here $\varphi \leftrightarrow \psi$ is an abbreviation for $(\varphi \rightarrow \psi) \wedge (\psi \rightarrow \varphi)$.

1. $\neg\varphi \leftrightarrow \neg\neg\neg\varphi$
2. $(\varphi \wedge \neg\psi) \rightarrow \neg(\varphi \rightarrow \psi)$
3. $(\varphi \rightarrow \psi) \rightarrow (\neg\neg\varphi \rightarrow \neg\neg\psi)$
4. $(\neg\neg(\varphi \rightarrow \psi) \leftrightarrow (\neg\neg\varphi \rightarrow \neg\neg\psi)$
5. $\neg\neg(\varphi \wedge \psi) \leftrightarrow (\neg\neg\varphi \wedge \neg\neg\psi)$
6. $\neg\neg(\forall x)\varphi(x) \rightarrow (\forall x)\neg\neg\varphi(x)$

DISJUNCTION AND EXISTENCE PROPERTIES.

The frame definition of intuitionistic validity makes it remarkably simple to prove the existence and disjunction properties.

THEOREM (DISJUNCTION PROPERTY). Let L be an intuitionistic predicate logic without function symbols. Then L has the disjunction property, that is $(\varphi_1 \vee \varphi_2)$ in L intuitionistically valid implies that one of φ_1, φ_2 is intuitionistically valid.

PROOF. Recall our assumption that L has at least one constant. Look at the contrapositive of the disjunction property, that is suppose φ_1 is not forced by p_1 in a frame $\mathcal{F}_1$ with partially ordered set P_1 and φ_2 is not forced by p_2 in another frame $\mathcal{F}_2$ with partially ordered set P_2.

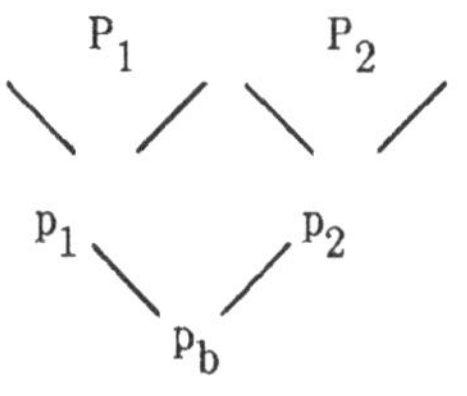

P

By the restriction lemma we may assume p_1 is least in P_1, p_2 is least in P_2. Make P_1 and P_2 and the sets of constants involved disjoint. Let P be the union of P_1, P_2, and $\{p_b\}$, with p_b new. Make P into a partial order by ordering P_1, P_2 as before and putting p_b below p_1 and p_2 to get a partial order P. Let C(p), A(p) be defined for p in P_1 or in P_2 as they were in the original frames. Let $C(p_b)$ consist of all constants in the language, non–empty by assumption, let $A(p_b)$ be the empty set. In this frame since p_b does not force φ_1, because p_1 extends p_b (monotonicity lemma) and p_b does not force φ_2 because p_2 extends p_b, we conclude from the definition of forcing that p_b does not force $\varphi_1 \vee \varphi_2$, contrary to hypothesis.

THEOREM (EXISTENCE PROPERTY). Let L be an intuitionistic predicate logic with no function symbols and at least one constant. Then L has the existence property, that is if $(\exists x)\varphi(x)$ is an intuitionistically valid statement in L, then for some c in L, $\varphi(c)$ is intuitionistically valid.

PROOF. Suppose for no constant c in L is $\varphi(c)$ intuitionistically valid. Then for each c there is an L–frame with partially ordered set P_c and element p_c not forcing $\varphi(c)$. Without loss of generality by the restriction lemma, p_c may be made the least element of P_c and all the P_c's may be made disjoint. Take the union of all P_c and the union of the partial orders and add a new bottom element p_b under all the p_c to get a partially ordered set P. We use P to construct an L–frame. Let $C(p_b)$ be the set of all constants in L and $A(p_b)$ be empty. Imitate the argument above. Since $(\exists x)\varphi(x)$ is forced by all p in all frames, it is forced by p_b. By the definition of forcing, for some c in L, p_b forces $\varphi(c)$. By the monotonicity lemma, p_c forces $\varphi(c)$, contrary to hypothesis.

EXERCISE. Let K be the set of constants occurring in $(\exists x)\varphi(x)$ and suppose that $(\exists x)\varphi(x)$ is intuitionistically valid. Show that if K is non–empty, then for some c in K, $\varphi(c)$ is intuitionistically valid. In case K is empty, show that $\varphi(c)$ is intuitionistically valid for any constant c.

SEMANTIC CONSEQUENCE

DEFINITION. Suppose a predicate logic L is specified and φ, $\psi_1, \ldots, \psi_k$ are statements in L. Call φ a semantic consequence of $\psi_1, \ldots, \psi_k$ if for any L–frame and any p, if p forces $\psi_1, \ldots, \psi_k$, then p forces φ.

Semantic consequence in this sense reduces to intuitionistic validity

LEMMA. φ is a semantic consequence of $\psi_1, \ldots, \psi_n$ if and only if $\psi_1 \wedge \ldots \wedge \psi_n \to \varphi$ is intuitionistically valid.

PROOF. Suppose φ is a semantic consequence of $\psi_1, \ldots, \psi_n$. We prove that if $\mathcal{F}$ is a frame and p is in $\mathcal{F}$, then p forces $\psi_1 \wedge \ldots \wedge \psi_n \to \varphi$. For this, suppose $p' \geq p$ forces $\psi_1 \wedge \ldots, \wedge \psi_n$ in $\mathcal{F}$. Then p' forces $\psi_1, \ldots, \psi_n$ by the definition of forcing of $\wedge$. By the definition of semantic consequence, p' forces φ in $\mathcal{F}$. This says p forces $\psi_1 \wedge \ldots \wedge \psi_n \to \varphi$ in $\mathcal{F}$. Conversely suppose $\psi_1 \wedge \ldots \wedge \psi_n \to \varphi$ is forced by all p in all L–frames. Suppose p forces $\psi_1, \ldots, \psi_n$ in a frame $\mathcal{F}$. By the definition of forcing, p forces $\psi_1 \wedge \ldots \wedge \psi_n$. Hence p forces φ in $\mathcal{F}$. So φ is a semantic consequence of $\psi_1, \ldots, \psi_n$.

REMARK. The phrase "in any L–frame" can be replaced by "in any frame in a class K " and the lemma still holds.

This lemma says that at least for finite sets of axioms, it is possible to develop their semantic consequences using the semantic notion of validity in frames without further apparatus. (This is also true for infinite sets of premises. This is a corollary to the systematic tableaux method when used for deductions. It also follows by a later coding of frames into classical models.)

CONGRUENCE AND APARTNESS

Equality in intuitionistic systems requires extensive discussion. We use a distinction made over two hundred years ago by Lagrange to motivate the discussion. In papers in 1772 Lagrange distinguished between two kinds of algebra and wrote a seminal paper on each. One leads to a definition of congruence that is suitable for constructive algebra. The other leads to a subtle refinement of congruence, apartness, used by Brouwer for his real number theory.

Congruence. The first paper of Lagrange dealt with algebra as a calculus of symbolic manipulation of strings of symbols. Over a hundred subsequent years this tradition was developed by such figures as Gauss, Cauchy, Galois, Jacobi, Kummer, Kronecker (1882). Assuming the positive integers as known, Kronecker gave computational constructions for the rational integers, the rational numbers, the integers mod p, finite dimensional polynomial domains over an already constructed ring, quotients of polynomial rings already constructed modulo finitely generated ideals. This was computational algebraic number theory and geometry, and his treatment was called elimination theory. This subject is outlined in small print in Van Der Waerden's Modern Algebra for the first several editions. In recent times, this subject has been made practical for computation by Gröbner and Buchburger, and is an active research field.

Here are the axioms assuring that $=$ is a congruence relation. Write $c = d$ instead of $=(c, d)$. We emphasize that in both classical models and frames, we definitely allow "=" to denote a congruence relation rather than merely identity. Thus if we speak of the integers mod 2, $\mathbb{Z}$ (mod 2), we allow the domain to be $\mathbb{Z}$ with the congruence relation $x = y$ if $x-y$ is divisible by 2. We do not wish to deal with the corresponding equivalence classes.

i) $(\forall x)(x = x)$
ii) $(\forall x)(\forall y)(x = y \rightarrow y = x)$
iii) $(\forall x)(\forall y)(\forall z)(x = y \wedge y=z \rightarrow x = z)$.
iv) $(\forall x_1)...(\forall x_n)(\forall y_1)...(\forall y_n)((x_1 = y_1 \wedge ... \wedge x_n = y_n \wedge R(x_1,...,x_n)) \rightarrow R(y_1,...,y_n))$

for each relation symbol R in the language.

If we wish to use frames to discuss partial knowledge of such an congruence in constructive algebra of the kind referred to above, we must allow congruence relations on $A(p)$ which may be arbitrary equivalence relations rather than only the identity relation. Here is why. Define a congruence to be discrete if

$$(\forall x)(\forall y)(x = y \vee \neg(x = y))$$

Notice this is the law of the excluded middle for congruence.

LEMMA. In a frame suppose p forces the congruence axioms. If for all $p' \geq p$, the congruence $=$ in $C(p')$ is identity, then p forces $=$ to be discrete.

PROOF. If for all $p' \geq p$, the equality $=$ in $C(p')$ is identity, then for all $p' \geq p$,
i) Identical constants c, d have $c = d$ in $A(p')$, so p' forces $c = d$.
ii) No two distinct constants c, d in $C(p')$ have $c = d$ in $A(p'')$ for any $p'' \geq p$. So for any two such distinct constants c, d, p' forces $\neg(c = d)$

This says p forces $=$ to be discrete. That is, if we restrict ourselves to the identity relation as interpretation of "=", then we are restricting ourselves to discrete congruence relations, called decidable equalities in intuitionist tradition. This turns out to be inappropriate for the real numbers, for example. See the discussion below.

Apartness. The second paper of Lagrange in 1772 was the first really comprehensive paper on numerical solution of algebraic equations with real coefficients. He makes it clear that to solve an equation is to give a method of computing closer and closer approximations to a solution from closer and closer approximations to the coefficients. He recognizes that this is a quite different situation than that above, the real number coefficients are not merely treated as formal strings of symbols. The whole tradition of numerical analysis from Lagrange to the present day is derived from this point of view. This is the tradition of which Brouwer is the intellectual heir. Brouwer's treatment of equality of reals is based on the notion of apartness, that is x and y are apart (written here $x||y$) if there is an integer n with $x-y > 1/n$. The point of this approach only comes into view when a computational point of view is taken, see the remark below the definition. When apartness is desired as primitive, $x = y$ should be defined as $\neg(x \mid\mid y)$. Thus x, y are equal in a state of knowledge if no future state of knowledge witnesses their being apart. This is the first axiom below.

DEFINITION. An apartness relation $\mid\mid$ is a binary relation such that

1) $(\forall x)(\forall y)(x = y \leftrightarrow \neg(x \mid\mid y))$
2) $(\forall x)(\forall y)(x \mid\mid y \rightarrow y \mid\mid x)$
3) $(\forall x)(\forall y)(\forall z)(x \mid\mid y \rightarrow x \mid\mid z \vee y \mid\mid z)$
4) Also the congruence axioms with $\neg(x \mid\mid y)$ substituted for $=$.

REMARK. Observe that the first axiom does not say $x \mid\mid y$ is the same thing as $\neg(x = y)$.

The rest of this section assumes the reader has some knowledge of Turing machines. If not, the reader should proceed to the next section. Here are some explanations of why apartness plays a role as a separate notion in constructive foundations of mathematics. Think of each real number as given by a real number generator, a pair consisting of a Cauchy sequence and a rate of convergence function for that Cauchy sequence. This is the line of reasoning employed by Bishop [1967].

DEFINITION. Define a real number generator as a pair (f, g), where f is a sequence of rationals $f(n) = r_n$ and g is a function with integer arguments and values such that for all N, $|r_n - r_m| < 1/N$ for $m, n \geq g(N)$. – i. e., $^1/_N$ is an ϵ for which $g(N)$ supplies a δ. Such a g is called a rate of convergence function.

Use as a model for computing an input–output Turing machine with an auxiliary work tape (see Hopcroft and Ullman [1979]). A real number generator is to be encoded on a tape by all triples $(0, f(0), g(0))$, $(1, f(1), g(1))$, ..., in any order. A little thought shows that a Turing machine can be built which, applied to a pair of real number generators encoded on input tapes, stops if and only if those generators define distinct reals. This is because if reals are apart, that is there is an n with their difference at least $1/n$, then this can be detected from the sequences and their rates–of–convergence functions in a finite length of time. But there is no Turing machine which, applied to a pair of real number generators written on input tapes, stops if and only if they are the same. A finite part of the sequences and their rates–of–convergence functions gives no such information. This indicates that apartness on the reals is better behaved from a constructive point of view than equality, since apartness can be detected and equality cannot. Brouwer chose apartness as fundamental for discussing reals. Apartness was axiomatized by Heyting.

EXERCISE. Verify the assertions made about Turing machines.

Functions. If we wish to discuss ordinary mathematical systems it is necessary to be able to treat functions conveniently. The definition we adopt is that functions of n variables are n+1–ary relations which are single valued and total.

(SINGLE VALUED) $(\forall x_1)\ldots(\forall x_n)(\forall x)(\forall y)(R(x_1,\ldots,x_n, x) \wedge R(x_1,\ldots, x_n,y) \rightarrow x = y)$

(TOTAL) $(\forall x_1)\ldots(\forall x_n)(\exists x)R(x_1,\ldots,x_n, x)$

If p forces these axioms, then in the associated classical model M(p), R defines a function f of n variables on the cartesian product $M(p)^n$ to M(p), but only relative to the congruence relation denoted by = on M(p).

A wider treatment of the concept of function would encompass functions which are not defined on some x in M(p) but which are defined for the same x in M(q) for some $q \geq p$. This might use $\neg(\forall x)\neg$ instead of $(\exists x)$ in the definitions above.

Groups and Rings. Groups, rings, and fields are universally studied algebraic structures. These structures are ubiquitous in pure and applied mathematics. Prior to the 1870's their theory was constructive (Dedekind's ideal theory was an exception, but Kronecker developed the same theorems at about the same time constructively). This constructive tradition includes the works of Lagrange (1770's), Cauchy (1815), Abel and Galois (1820's), Kummer (1840's), Jordan (1860's), Kronecker (1880's).

What definitions can be given in frames for the notions of intuitionistic groups, rings, fields?

We give the axioms for groups with • as group operation and e as identity, but think of •(x,y,z) as a ternary relation x•y = z. This is because of subtleties in the use of function symbols in frames, which we do not wish to consider here. The group axioms are the axioms for congruence and those axioms guaranteeing that • is a function.

EXISTENCE OF IDENTITY.
$(\forall x)(\bullet(x,e,x) \wedge \bullet(e,x,x))$

EXISTENCE OF INVERSE.
$(\forall x)(\exists y)(\bullet(x,y,e) \wedge \bullet(y,x,e))$

ASSOCIATIVE LAW.
$(\forall x)(\forall y)(\forall z)((\forall w)(\forall u)(\forall a)(\bullet(x,y,u) \wedge \bullet(u,z,a) \wedge\bullet(y,z,w) \rightarrow \bullet(x,w,a))$

Note that the interpretation of $\forall$, $\exists$ in frames means that if p forces the group axioms, then for all $p' \geq p$, C(p') is a group relative to the congruence relation denoted by =. We explicitly allow a congruence relation on the underlying set in the definition of group. We call a frame for a language with ternary relation $\bullet$, binary relation $=$, constant e an intuitionistic group if every p forces the group axioms listed above. Thus with each p in P is associated a group H_p. As we go from p to q, $p \leq q$, we can add elements to C(p) get a larger group C(q) and also reduce by a homomorphic image at the same time. We get a collection of groups $G_p = C(p)$ indexed by a partially ordered set P such that whenever $p \leq q$, we have a homomorphism H_{pq} of G_p into G_q such that H_{pp} is the identity and $H_{pq}H_{qr} = H_{pr}$. Conversely, any such category of groups so arises from a frame as described above. So such a category of groups will serve as an intuitionistic group. The discussion is identical for the additional axioms for a commutative ring with unit. This now allows us to give examples.

EXAMPLE. Let $\mathbb{Z}$ be the ring of integers with identity as the congruence, let $\mathbb{Z}$ (mod 2) have as domain the set $\mathbb{Z}$ of integers with congruence mod 2 as the interpretation of =. Then, using the obvious maps, we have a frame

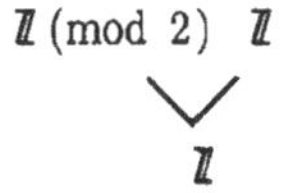

Call the lower node 0, the upper left node 00, the right upper node 01 to establish the partial order. The bottom node 0 forces the commutative ring with unit axioms. This ring is not discrete, that is 0 does not force $(\forall x)(x = 0 \vee \neg(x = 0))$, since 0 does not force $2 = 0$ and 00 does force $2 = 0$. 0 does not force "every non–zero element has an inverse" and does not force "there is a non–zero non–invertible element".

EXAMPLE. Let $\mathbb{Q}$ be the ring of rational numbers, let $\mathbb{Q}(\sqrt{2})$ be all $a + b\sqrt{2}$ with a, b rational. Use the natural embedding $\mathbb{Q}$ into $\mathbb{Q}(\sqrt{2})$ and the identity as congruence on each (regard them as having discrete equalities).

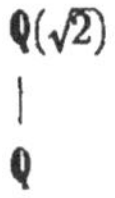

Call the lower node 0, the upper 00 to establish the partially ordered set P. Then 0 forces the field axioms and discrete equality but 0 does not force "there is a square root of 2", and does not force "there is not a square root of 2".

EXAMPLE. Here is another way to motivate frame groups. One traditional way of dealing with arbitrary groups is to think of a group G as given by a set H of generators and set R of relations. In this conception a group G consists of words (that is, strings $y_1 \bullet y_2 \bullet \ldots y_n$ where each y_i is an h in H or h^{-1} with h in H) and a congruence relation "=" (that is, $w_1 = w_2$ iff w_1 can be transformed into w_2 by a finite number of applications of the group laws and usual properties of congruence.) Thus many different words are names of the same abstract group element. From a constructive point of view, should we divide out and immediately think of the group as a set of congruence classes of words under =, as is done in classical mathematics with classical logic? Suppose even that the group is finitely presented, that is H and R are finite. Then we can effectively generate all pairs (x, y) of equal words by systematically using the group rules repeatedly. The word problem for groups is to determine, if a group is finitely presented, whether or not there is an algorithm to decide if a pair of words r, s represent the same group element, that is under the group congruence relation =, whether $w_1 = w_2$ or $\neg(w_1 = w_2)$. If there are no relations present, that is, if we are in a free group, there is an easy algorithm for equality of words, each group element has a canonical form. But in 1954 finitely presented groups were exhibited for which there is no such algorithm (unsolvability of the word problem for groups) by Novikoff and by Boone. This means that even for a finitely presented group, all we can expect is for a variety of mathematicians to discover gradually more and more group equalities and inequalities, but never all inequalities by any fixed procedure. This is a hint that we don't want the law of the excluded middle assumed for equality. We may have to deal with names of objects, and may not be able to determine whether or not names denote the same classical object.

Historical Note. The word problem for finitely presented groups arose directly out of elementary topology problems pursued by Dehn in the first decade of the century. At that time the notion of homotopic paths was known, and finding the first homotopy group of a space was a problem of considerable importance. For manifolds arising classically, the problem was reduced to calculations on generators and relations, in fact every finitely presented group was shown by Dehn to arise as the first homotopy group of a suitable two dimensional complex. He could write down generators and relations, but whether two words were equal on that basis he could not determine. This was frustrating, because this means exactly that he could not tell whether a path was homotopic to the identity. Now we know there is no such algorithmic test. Now let's move over to frames. Think of a group G as being defined by generators and relations. Assume we know at all times all the group laws in the free group generated by the generators thus far given. Suppose the generators and relations are gradually discovered by one of several investigators and we don't know exactly which of the generators and relations will be discovered next by which investigator, but after they are discovered everybody knows them. Suppose, due to their mental limitations, there are constraints on what discoveries in the form of generators and relations each investigator can make next, based on what is known. Then the possible "states of knowledge" about the group based on its generators and relations form a partially ordered set P where $p \leq q$ means that state of knowledge q has at least as many generators and relations but possibly more are known than at p. Then a frame should assign to p a free group and a congruence relation corresponding to a set of relations. If $p \leq q$, the free group G_q assigned to q contains the free group G_p assigned to p, the congruence relation $=_q$ on G_q contains the congruence relation $=_p$ on G_p. But we have seen that all these requirements on the frames simply arise when the group axioms are written in relational form.

INTUITIONISTIC TABLEAUX

There are many possible variants on the tableaux method suitable for intuitionistic propositional and predicate logic due to Kripke, Hughes and Cresswell, Fitting, and others. The one we choose is designed to match the definition of frame exactly, so that the systematic tableaux will represent exactly a systematic search for a frame all of whose p force formula φ . It is a variant of Hughes and Cresswell's, or Fitting's prefixed tableaux. Had a variant definition of frame been chosen for this exposition (there are many), we would have modified the definition of tableaux to match. We emphasize that this matching of definitions of tableaux and frame is always possible. The definition of tableaux we adopt is not the most succinct, but at least it holds no mysteries.

Labels for intuitionistic tableaux will be forcing assertions of the form $Tp\Vdash\varphi$ and $Fp\Vdash\varphi$ for statements φ, where p is from a partially ordered set P. We read $Tp\Vdash\varphi$ as p forces φ, $Fp\Vdash\varphi$ as p doesn't force φ. A branch b is open if it does not have for any statement φ and partial order element q both $Tq\Vdash\varphi$ and $Fq\Vdash\psi$ as entries on b. If a branch is not open, it is declared <u>closed</u>. As soon as a branch is found to be closed, it is marked with an "x" and not developed further. The rules for developing an intuitionistic tableaux given below reflect exactly all constructions necessary to prove completeness.

Atomic Tableaux.

<u>OR</u>

1.

$$\begin{array}{ccc} & Tp\Vdash\varphi \vee \psi & \\ \swarrow & & \searrow \\ Tp\Vdash\varphi & & Tp\Vdash\psi \end{array}$$

2.

$$\begin{array}{c} Fp\Vdash\varphi \vee \psi \\ | \\ Fp\Vdash\varphi \\ | \\ Fp\Vdash\psi \end{array}$$

<u>AND</u>

3.

$$Tp \Vdash \varphi \wedge \psi$$
$$Tp \Vdash \varphi$$
$$Tp \Vdash \psi$$

4.

$$Fp \Vdash \varphi \wedge \psi$$
$$Fp \Vdash \varphi \qquad Fp \Vdash \psi$$

IMPLIES

5.

$$Tp \Vdash \varphi \rightarrow \psi$$
$$Fp' \Vdash \varphi \qquad Tp' \Vdash \psi$$
for any $p' \geq p$

6.

$$Fp \Vdash \varphi \rightarrow \psi$$
$$Tp' \Vdash \varphi$$
$$Fp' \Vdash \psi$$
for some new $p' \geq p$

NOT

7.

$$Tp \Vdash \neg\varphi$$
$$Fp' \Vdash \varphi$$
for any $p' \geq p$

8.

$$Fp \Vdash \neg\varphi$$
$$Tp' \Vdash \varphi$$
for some new $p' \geq p$

THERE EXISTS

9.

$$Tp \Vdash (\exists x)\varphi(x)$$
$$Tp \Vdash \varphi(c)$$
for some new c

10.

$$Fp \Vdash (\exists x)\varphi(x)$$
$$Fp \Vdash \varphi(c)$$
for any c

FOR ALL

11.
$$\begin{array}{c} Tp\Vdash(\forall x)\varphi(x) \\ | \\ Tp'\Vdash\varphi(c) \end{array}$$
for any $p' \geq p$, any c

12.
$$\begin{array}{c} Fp\Vdash(\forall x)\varphi(x) \\ | \\ Fp'\Vdash\varphi(c) \end{array}$$
for some new $p' \geq p$, for some new c

FALSE

13.
$$\begin{array}{c} Tp\Vdash F \\ | \\ x \end{array}$$

This last will not be often used: only if we choose to bring the F symbol into the theory itself.

REMARK. The monotonicity lemma can be turned into a tableaux rule and used to shorten proofs by making contradictions appear sooner. But monotonicity does not correspond exactly to one of the clauses in the definition of forcing in frames and is not used in the completeness proof. Whenever a rule requiring the introduction of a new extension p' is applied to extend a branch b, (rules 5, 6, 7, 8, 11, 12), monotonicity can be applied repeatedly to append to the branch b all $Tp'\Vdash\varphi$ for which there is a $p \leq p'$ such that $Tp\Vdash\varphi$ is already on b. This is called the branch modification rule by Fitting.

MONOTONICITY

13.
$$\begin{array}{c} Tp\Vdash\varphi \\ | \\ Tp'\Vdash\varphi \end{array}$$
for any $p' \geq p$

With monotonicity present the rule for $Tp\Vdash(\forall x)\varphi(x)$ can be made
$$\begin{array}{c} Tp\Vdash(\forall x)\varphi(x) \\ | \\ Tp\Vdash\varphi(c) \end{array}$$
any c

because for $p' > p$, $Tp'\Vdash\varphi(c)$ can be introduced by monotonicity.

REMARK. "new c" means c does not occur on the branch being developed, "any c" means any c occurring already on the branch being developed, " **NEW** $p' > p$ " means a p' not on the branch as yet and incomparable with all q not $< p$ and already on the branch, "any p" means any p already on the branch being developed.

REMARK. The definition of a tableaux as constructed by these rules is like that for classical logic in Smullyan [1968]. In the formal definition we give, a **WHOLE** atomic tableaux is appended at the base of a branch when the branch is extended, but in the examples the apex of each tableaux is omitted. Instead the lines of the tableaux are numbered to the left. To the right of each entry is listed the number of the omitted apex.

REMARK. Smullyan's classical logic tableaux [1968] in fact result when only the one element partially ordered set $\{0\}$ is ever allowed, and $T0\Vdash\varphi$ is written $T\varphi$, $F0\Vdash\varphi$ is written $F\varphi$.

CORRECTNESS THEOREM. Suppose $\mathcal{F} = (P, \leq, A(p), C(p))$ is a frame and Σ is a tableaux. Suppose either
$Fp\Vdash\varphi$ is the apex of tableaux Σ and p does not force φ in $\mathcal{F}$, or
$Tp\Vdash\varphi$ is the apex of Σ and p does force φ in $\mathcal{F}$.

Then there is a branch b through the tableaux and
1) an assignment of values in P to all partial order element constants q occurring in signed formulas on b,
2) an assignment of values in C(q) to all domain constants c occurring in signed formulas $Tq\Vdash\psi$ or $Fq\Vdash\psi$ on b
such that

$Tq\Vdash\psi$ on b implies q forces ψ in $\mathcal{F}$, and
$Fq\Vdash\psi$ on b implies q does not force ψ in $\mathcal{F}$.

Thus, the theorem asserts: if a frame agrees with the apex of a tableaux, it agrees with a whole path of it. An immediate consequence is the traditional statement of the correctness, or soundness theorem: if there is a tableaux proof of φ then φ is true in every frame.

PROOF. The proof is by induction on the construction of a tableaux by using the twelve development rules. The inductive hypothesis is that for a given tableaux Σ there is a branch b through the tableaux Σ and a pair of assignments for b which agree with forcing on the frame. For the induction we must show that if the tableaux Σ is extended to a tableaux Σ' by applying one of the twelve rules for decoding logical connectives, then there is an extension of b to a b' through Σ' and two assignments for b' which agree with forcing on the frame. Tableaux rules are invented to make this happen. For a typical case suppose the tableaux rule applied to Σ to append an entry to the base of the b given by the inductive hypothesis is an application of rule 12. So $Fp\Vdash(\forall x)\varphi(x)$ is on the tableaux branch b and to get Σ' we append to the base of b the atomic tableaux below.

$$
\begin{array}{l}
Fp\Vdash(\forall x)\varphi(x) \\
\quad | \\
Fp'\Vdash\varphi(c)
\end{array}
$$

By inductive hypothesis there is a pair of assignments for b making $p\Vdash(\forall x)\varphi(x)$ true. The definition of forcing says we can find a $p' \geq p$ and a c in $C(p')$ such that $p'\Vdash\varphi(c)$. This shows how to extend the pair of assignments to a b' through Σ'. □

EXERCISE. Carry out the other eleven cases of the proof of correctness.

We show how construction of a tableaux for propositional logic yields a counterexample as easily as a proof. (This is discussed in more detail in the section on a decision method for intuitionistic propositional logic.).

EXAMPLE 11. Consider $F0\Vdash A \to (A \to B)$

1	$F0\Vdash A \to (A \to B)$	
2	$T00\Vdash A$	by 1
3	$F00\Vdash(A \to B)$	by 1
4	$T000\Vdash A$	by 3
5	$F000\Vdash B$	by 3

Here 1, a false implication, is assumed to not be forced at state 0. Using the tableaux rule for implication, we introduce a larger state 00 with 00 forcing antecedent A in line 2 and 00 not forcing the consequent $(A \to B)$ in line 3. Using the tableaux rule for implication on line 3, we introduce a larger state 000 and in line 4 assert that 000 forces A and in line 5 that 000 does not force B. We stop at 000. Why? The only rule that can be used now is the monotonicity rule, which will allow larger states to be brought in, such as $T0000\Vdash A$, $T000000\Vdash A$, etc, after which the other rules could possibly be applied. But these partial order elements if introduced would force the same true statements as one of 0, 00, 000, and would have one of these as an initial segment. For forming a frame there is no point in introducing new partial order elements which force the same true statements as smaller partial order elements. So these new elements are not needed, and in fact 000 forces the same φ as 00 and is not needed in a frame counterexample to $A \to (A \to B)$. The fact that 000 forces the same things as 00 is our test that we do not need to continue the tableaux. This gives rise (looking at the forced true statements) to the frame

A

Let 0 be the lower node, 00 the upper node. In this model 0 forces nothing, 00 forces only A. Then 0 does not force $A \rightarrow (A \rightarrow B)$.

EXAMPLE 12. Consider $\varphi \vee \neg\varphi$.

1	$F0 \Vdash \varphi \vee \neg\varphi$	
2	$F0 \Vdash \varphi$	by 1
3	$F0 \Vdash \neg\varphi$	by 1
4	$T00 \Vdash \varphi$	by 3

Note that now only the monotonicity rule can be applied, and it yields only the same forced true statements $T000 \Vdash \varphi$, $T0000 \Vdash \varphi$, etc. So for constructing a counterexample (or constructing a proof), nothing new will be obtained by continuing. Letting $\mathbf{A}(0)$ be empty, $\mathbf{A}(00)$ be $\{\varphi\}$, we get a frame with $\varphi \vee \neg\varphi$ not forced. This is the same frame as in an example above, but obtained more systematically.

EXAMPLE 13. Consider $(\neg\varphi \rightarrow \neg\psi) \rightarrow (\psi \rightarrow \varphi)$.

1	$F0 \Vdash (\neg\varphi \rightarrow \neg\psi) \rightarrow (\psi \rightarrow \varphi)$		
2	$T00 \Vdash \neg\varphi \rightarrow \neg\psi$		by 1
3	$F00 \Vdash \psi \rightarrow \varphi$		by 1
4	$T000 \Vdash \psi$		by 3
5	$F000 \Vdash \varphi$		by 3
6	$T000 \Vdash \neg\varphi \rightarrow \neg\psi$		by 2
7	$F000 \Vdash \neg\varphi$	$T000 \Vdash \neg\psi$	by 6
8	$T0000 \Vdash \varphi$	$F000 \Vdash \psi$	by 7
		x by 4	

Note that no new k with $Tp \Vdash k$ will turn up on further development, so letting 0, 00 force no atomic statements and $000 \Vdash \psi$ and $0000 \Vdash \varphi$ will give the desired counterexample frame. But 0, 00 might as well be collapsed to one as far as forcing goes. So we end up with 0, 00, 000 as the partially ordered set, $\mathbf{A}(0) =$ null set, $\mathbf{A}(00) = \{\psi\}$, $\mathbf{A}(000) = \{\varphi, \psi\}$.

EXAMPLE 14. Consider $A \rightarrow (B \rightarrow A)$.

1 $F0 \Vdash A \rightarrow (B \rightarrow A)$

2 $T00 \Vdash A$ by 1

3 $F00 \Vdash B \rightarrow A$ by 1

4 $T000 \Vdash B$ by 3

5 $F000 \Vdash A$ by 3

6 $T000 \Vdash A$ by 2

x

Here 2, $T00 \Vdash A$, is introduced by a false implication, 5, $F000 \Vdash A$, is introduced by a false implication, 6, $T000 \Vdash A$, is introduced by monotonicity using 2. This gives a direct contradiction, so $A \rightarrow (B \rightarrow A)$ is intuitionistically valid and this is a tableaux proof of $A \rightarrow (B \rightarrow A)$.

EXAMPLE 15. Consider $(\exists x)(\varphi(x) \vee \psi(x)) \rightarrow (\exists x)\psi(x) \vee (\exists x)\psi(x)$

1 $F0 \Vdash (\exists x)(\varphi(x) \vee \psi(x)) \rightarrow (\exists x)\psi(x) \vee (\exists x)\psi(x)$

2 $T00 \Vdash (\exists x)(\varphi(x) \vee \psi(x))$ by 1

3 $F00 \Vdash (\exists x)\varphi(x) \vee (\exists x)(\psi(x)$ by 1

4 $T00 \Vdash \varphi(c) \vee \psi(c)$ by 2

5 $F00 \Vdash (\exists x)\varphi(x)$ by 3

6 $F00 \Vdash (\exists x)\psi(x)$ by 3

7 $F00 \Vdash \varphi(c)$ by 5

8 $F00 \Vdash \psi(c)$ by 6

9 $T00 \Vdash \varphi(c)$ | $T00 \Vdash \psi(c)$ by 4

x | x by 7, 8

So $(\exists x)(\varphi(x) \vee \psi(x)) \rightarrow (\exists x)\psi(x) \vee (\exists x)\psi(x)$ is intuitionistically valid, and this is a tableaux proof.

EXAMPLE 16. Consider $(\forall x)(\varphi(x) \wedge \psi(x)) \rightarrow (\forall x)\varphi(x) \wedge (\forall x)\psi(x)$.

1	$F0 \Vdash (\forall x)(\varphi(x) \wedge \psi(x)) \rightarrow (\forall x)\varphi(x) \wedge (\forall x)\psi(x)$		
2	$T00 \Vdash (\forall x)(\varphi(x) \wedge \psi(x))$		by 1
3	$F00 \Vdash (\forall x)\varphi(x) \wedge (\forall x)\varphi(x)$		by 2
4	$F00 \Vdash (\forall x)\varphi(x)$	$F00 \Vdash (\forall x)\psi(x)$	by 3
5	$F000 \Vdash \varphi(c)$	$F000 \Vdash \psi(d)$	by 4
6	$T000 \Vdash (\forall x)(\varphi(x) \wedge \psi(x))$	$T000 \Vdash (\forall x)(\varphi(x) \wedge \psi(x))$	by 2
7	$T000 \Vdash \varphi(c) \wedge \psi(c)$	$T000 \Vdash \varphi(d) \wedge \psi(d)$	by 6
8	$T000 \Vdash \varphi(c)$	$T000 \Vdash \psi(d)$	by 5, 7 by 7
	x	x	

This is a tableaux proof of $(\forall x)(\varphi(x) \wedge \psi(x)) \rightarrow (\forall x)\varphi(x) \wedge (\forall x)\psi(x)$.
Note that 000 was used on two branches which have nothing to do with each other.

EXAMPLE 17. Consider $(\varphi \rightarrow \psi) \vee (\psi \rightarrow \varphi)$

1	$F0 \Vdash (\varphi \rightarrow \psi) \vee (\psi \rightarrow \varphi)$	
2	$F0 \Vdash \varphi \rightarrow \psi$	by 1
3	$F0 \Vdash \psi \rightarrow \varphi$	by 1
4	$T00 \Vdash \varphi$	by 2
5	$F00 \Vdash \psi$	by 2
6	$T01 \Vdash \psi$	by 3
7	$F01 \Vdash \varphi$	by 3

Observe that this is the first example in which the "new" $p' \geq p$ stipulation of rule 6 (applied here to line 3 to obtain lines 6 and 7) forces our frame to branch. Node 01 in line 6 was chosen as the least node greater than 0 incomparable with every p on the tree not ≤ 0. In fact, no linear (non–branching) frame can fail to force $(\varphi \rightarrow \psi) \vee (\psi \rightarrow \varphi)$ (exercise).

EXERCISE SET 2.

Construct tableaux starting with $F0 \Vdash \varphi$ to see that the following classically valid formulas are also intuitionistically valid. (Take $\varphi \leftrightarrow \psi$ to be an abbreviation for $(\varphi \rightarrow \psi) \wedge (\psi \rightarrow \varphi)$).

PROPOSITIONAL LOGIC

DISTRIBUTIVE LATTICE LAWS

1. $(\varphi \vee \varphi) \leftrightarrow \varphi$

2. $(\varphi \wedge \varphi) \leftrightarrow \varphi$
3. $(\varphi \wedge \psi) \leftrightarrow (\psi \wedge \varphi)$
4 $(\varphi \vee \psi) \leftrightarrow (\psi \vee \varphi)$
5. $((\varphi \wedge \psi) \wedge \sigma) \leftrightarrow (\varphi \wedge (\psi \wedge \sigma))$
6. $((\varphi \vee \psi) \vee \sigma) \leftrightarrow (\varphi \vee (\psi \vee \sigma))$
7. $(\varphi \vee (\psi \wedge \sigma)) \leftrightarrow ((\varphi \vee \psi) \wedge (\varphi \vee \sigma))$
8. $(\varphi \wedge (\psi \vee \sigma)) \leftrightarrow ((\varphi \wedge \psi) \vee (\varphi \wedge \sigma))$

PURE IMPLICATION LAWS

9. $\varphi \rightarrow \varphi$
10. $\varphi \rightarrow (\psi \rightarrow \varphi)$
11. $(\varphi \rightarrow \psi) \rightarrow ((\psi \rightarrow \sigma) \rightarrow (\varphi \rightarrow \sigma))$

INTRODUCTION AND ELIMINATION OF $\wedge$

12 $((\varphi \rightarrow (\psi \rightarrow \sigma)) \rightarrow ((\varphi \wedge \psi) \rightarrow \sigma)$
13. $((\varphi \wedge \psi) \rightarrow \sigma) \rightarrow ((\varphi \rightarrow (\psi \rightarrow \sigma))$
14. $(\neg\varphi \vee \psi) \rightarrow (\varphi \rightarrow \psi)$

DEMORGAN'S LAWS

15. $\neg(\varphi \vee \psi) \leftrightarrow (\neg\varphi \wedge \neg\psi)$
16. $(\neg\varphi \vee \neg\psi) \rightarrow \neg(\varphi \wedge \psi)$

CONTRAPOSITIVE

17. $(\varphi \rightarrow \psi) \rightarrow (\neg\psi \rightarrow \neg\varphi)$

DOUBLE NEGATION

18. $\varphi \rightarrow \neg\neg\varphi$

CONTRADICTION

19. $F \leftrightarrow (\varphi \wedge \neg\varphi)$

PREDICATE LOGIC

DISTRIBUTIVE LAWS

20. $(\exists x)(\varphi(x) \vee \psi(x)) \leftrightarrow (\exists x)\varphi(x) \vee (\exists x)\psi(x)$
21 $((\forall x)(\varphi(x) \wedge \psi(x)) \leftrightarrow (\forall x)\varphi(x) \wedge (\forall x)\psi(x)$
22. $(\varphi \vee (\forall x)\psi(x)) \rightarrow (\forall x)(\varphi \vee \psi(x))$, x not free in φ
23. $(\varphi \wedge (\exists x)\psi(x)) \rightarrow (\exists x)(\varphi \wedge \psi(x))$, x not free in φ
24. $(\exists x)(\varphi \rightarrow \psi(x)) \rightarrow (\varphi \rightarrow (\exists x)\psi(x))$, x not free in φ
25. $(\exists x)(\varphi \wedge \psi(x)) \rightarrow (\varphi \wedge (\exists x)\psi(x))$, x not free in φ

DEMORGAN LAWS

26 $\neg(\exists x)\varphi(x) \rightarrow (\forall x)\neg\varphi(x)$
27. $(\forall x)\neg\varphi(x) \rightarrow \neg(\exists x)\varphi(x)$
28. $(\exists x)\neg\varphi(x) \rightarrow \neg(\forall x)\varphi(x)$
29. $(\exists x)(\varphi(x) \rightarrow \psi) \leftrightarrow ((\forall x)\varphi(x) \rightarrow \psi)$, x not free in ψ
30. $((\exists x)\varphi(x) \rightarrow \psi) \rightarrow (\forall x)(\varphi(x) \rightarrow \psi)$, x not free in ψ

EXERCISE SET 3.

Below is a list of classically valid formulas φ which are not intuitionistically valid. For each, start a tableaux with $F0\Vdash\varphi$, develop it enough to produce a frame in which φ is not forced.

PROPOSITIONAL LOGIC

1. $(\varphi \vee \neg\varphi)$
2. $(\neg\neg\varphi \to \varphi)$
3. $\neg(\varphi \wedge \psi) \to (\neg\varphi \vee \neg\psi)$.
4. $\neg\varphi \vee \neg\neg\varphi$
5. $(\varphi \to \psi) \vee (\psi \to \varphi)$
6. $(\neg\neg\varphi \to \varphi) \to (\varphi \vee \neg\varphi)$
7. $(\neg\varphi \to \neg\psi) \to (\psi \to \varphi)$
8. $(\varphi \to \psi) \to (\neg\varphi \vee \psi)$

PREDICATE LOGIC

9. $\neg(\forall x)\varphi(x) \to (\exists x)\neg\varphi(x)$
10. $(\forall x)\neg\neg\varphi(x) \to \neg\neg(\forall x)\varphi(x)$
11. $(\forall x)(\varphi \vee \psi(x)) \to (\varphi \vee (\forall x)\psi(x))$
12. $((\varphi \to (\exists x)\psi(x)) \to (\exists x)(\varphi \to \psi(x))$
13. $(\forall x)\varphi(x) \to \psi) \to (\exists x)(\varphi(x) \to \psi(x))$
14. $((\forall x)(\varphi(x) \vee \neg\, \varphi(x)) \wedge \neg\neg(\exists x)\varphi(x)\ ) \to (\exists x)\varphi(x)$

COMPLETENESS OF THE TABLEAUX METHOD

The reader should consult the classical logic complete systematic tableaux definitions given in Smullyan [1968]. Unlike the examples above, in the systematic tableaux procedure we append the **WHOLE** atomic tableaux (including the apex) to the base of a branch b when we extend b, rather than omitting the apex.

Where will the required partial order P for Σ come from? P will be chosen a subset of a fixed partially ordered set **P**. The crucial property of **P** needed is that given any finite number of incomparable elements $p_1, \ldots, p_n$ of **P**, there is a $p > p_1$ in **P** such that $p, p_2, \ldots, p_n$ are incomparable. Any partially ordered **P** with this property would do. We use for **P** the set of all finite sequences of non–negative integers, partially ordered by $p \leq q$ if q is an extension of p.

We define an increasing sequence of finite tableaux Σ_n and take $\Sigma = \cup\, \Sigma_n$, which is therefore possibly infinite. Σ_0 is just $F0\Vdash\varphi$. Σ_{n+1} is obtained from Σ_n by attacking exactly one entry of Σ_n to produce Σ_{n+1}. An exact order of attack of entries must be specified so as to eventually meet every possible requirement. We leave this to the reader.

1. If an occurrence of $Tp\Vdash\varphi \vee \psi$ is the entry of Σ_n currently attacked, for each open branch b through that occurrence append to Σ_n the atomic tableaux below.

$$\begin{array}{ccc} & Tp\Vdash\varphi \vee \psi & \\ \swarrow & & \searrow \\ Tp\Vdash\varphi & & Tp\Vdash\psi \end{array}$$

2. If an occurrence of $Fp\Vdash\varphi \vee \psi$ is the entry of Σ_n currently attacked, for each open branch b through that occurrence, append to Σ_n the atomic tableaux below.

$$\begin{array}{c} Fp\Vdash\varphi \vee \psi \\ | \\ Fp\Vdash\varphi \\ | \\ Fp\Vdash\varphi \end{array}$$

3. If an occurrence of $Tp\Vdash\varphi \wedge \psi$ is the entry of Σ_n currently attacked, for each open branch b through that occurrence, append the atomic tableaux below.

$$\begin{array}{c} Tp\Vdash\varphi \wedge \psi \\ | \\ Tp\Vdash\varphi \\ | \\ Tp\Vdash\psi \end{array}$$

4. If an occurrence of $Fp\Vdash\varphi \wedge \psi$ is the entry of Σ_n currently attacked, for each open branch b through that occurrence, append to b the atomic tableaux below.

$$\begin{array}{ccc} & Fp\Vdash\varphi \wedge \psi & \\ \swarrow & & \searrow \\ Fp\Vdash\varphi & & Fp\Vdash\psi \end{array}$$

5. If an occurrence of $Tp\Vdash \varphi \rightarrow \psi$ is the entry of Σ_n currently attacked, for each open branch b through that entry and each $p' \geq p$ on b append to b the atomic tableaux below.

$$\begin{array}{ccc} & Tp\Vdash\varphi \rightarrow \psi & \\ \swarrow & & \searrow \\ Fp'\Vdash\varphi & & Tp'\Vdash\psi \end{array}$$

6. If an occurrence of $Fp\Vdash\varphi \rightarrow \psi$ is the entry of Σ_n currently attacked, for each open branch b through that entry choose a $p' \geq p$ in **P** not on branch b and incomparable with each q on b which is not $\leq p$ and append to b the atomic tableaux below.

$$\begin{array}{c} Fp\Vdash\varphi \rightarrow \psi \\ | \\ Tp'\Vdash\varphi \\ | \\ Fp'\Vdash\psi \end{array}$$

7. If an occurrence of $Tp\Vdash\neg\varphi$ is the entry of Σ_n currently attacked, for each open branch b through that entry and each $p' \geq p$ occurring on b, append to b the atomic tableaux below.

$$\begin{array}{c} Tp\Vdash\neg\varphi \\ | \\ Fp'\Vdash\varphi \end{array}$$

8. If an occurrence of $Fp\Vdash\neg\varphi$ is the entry of Σ_n currently attacked, for each open branch b through that entry choose a $p' \geq p$ in **P** not on branch b and incomparable with each q on b not $\leq p$ and append to b the atomic tableaux below.

$$\begin{array}{c} Fp\Vdash\neg\varphi \\ | \\ Tp'\Vdash\varphi \end{array}$$

9. If an occurrence of $Tp\Vdash(\exists x)\varphi(x)$ is the entry of Σ_n currently attacked, for each open branch b through that occurrence chose a constant c not occurring on the branch and append the atomic tableaux below.

$$\begin{array}{c} Tp\Vdash(\exists x)\varphi(x) \\ | \\ Tp\Vdash\varphi(c) \end{array}$$

10. If an occurrence of $Fp\Vdash(\exists x)\varphi(x)$ is the entry of Σ_n currently attacked, for each open branch b through that entry and each constant c occurring in a formula $Tp\Vdash\psi$ or $Fp\Vdash\psi$ on b, append to b the atomic tableaux below.

$$\begin{array}{c} Fp\Vdash(\exists x)\varphi(x) \\ | \\ Fp\Vdash\varphi(c) \end{array}$$

11. If an occurrence of $Tp\Vdash(\forall x)\varphi(x)$ is the entry of Σ_n currently attacked, for each open branch b through that entry and each $p' \geq p$ occurring on the branch b and each constant c occurring in a formula of the form $Tp'\Vdash\psi$ or $Fp'\Vdash\psi$ on b, append to b the atomic tableaux below.

$$\begin{array}{c} Tp\Vdash(\forall x)\varphi(x) \\ | \\ Tp'\Vdash\varphi(c) \end{array}$$

12. If an occurrence of $Fp\Vdash(\forall x)\varphi(x)$ is the entry of Σ_n currently attacked, for each open branch b through that entry choose a $p' \geq p$ in **P** not on branch b and incomparable with each q on b such that not $q \leq p$, and append the atomic tableau below.

$$\begin{array}{c} Fp\Vdash(\forall x)\varphi(x) \\ | \\ Fp'\Vdash\varphi(c) \end{array}$$

EXERCISE. Give a strategy for attacking requirements so that the complete systematic tableaux has the twelve features outlined above. (You may use the argument for classical logic tableaux from Smullyan [1968] as a model.)

THEOREM. Suppose b is an open branch on the complete systematic tableaux Σ. Define a frame $\mathcal{K}_b = (P, \leq, A(p), C(p))$ where

P consists of all p with a $Tp\Vdash\varphi$ or $Fp\Vdash\varphi$ on b

A(p) consists of all atomic φ such that $Tq\Vdash\varphi$ occurs on b for some $q \leq p$,

C(p) consists of all constants c occurring in statements in A(p). Then

$Tp\Vdash\varphi$ on branch b implies p forces φ in $\mathcal{K}_b$,

$Fp\Vdash\psi$ on branch b implies p does not force ψ in $\mathcal{K}_b$.

Thus to each non–contradictory branch b of the tableaux there corresponds a frame $\mathcal{K}_b$ that agrees with every node.

PROOF. We divide the proof into one base step for atomic formulas and twelve induction steps, one for each of the twelve conditions in the definition of Σ.

BASE STEP.

0. Suppose $Tp\Vdash\varphi$ or $Fp\Vdash\psi$ is on open branch b with φ, ψ atomic. Then $Tp\Vdash\varphi$ on b and φ atomic imply that φ is in A(p), or that p forces φ in $\mathcal{K}_b$. If $Fp\Vdash\psi$ is on b, then $Tp\Vdash\varphi$ is not on b since b is open, so φ is not in A(p) and p does not force φ in $\mathcal{K}_b$.

INDUCTION STEPS. For an induction hypothesis, assume the theorem holds for shorter φ (and all p) than the one we are interested in. (We first treat the "easy" $\exists$, $\wedge$, $\vee$ clauses, namely those for which truth at node p depends only on the behavior at p)

1. Suppose that $Tp\Vdash\varphi \vee \psi$ is on b. Then either $Tp\Vdash\varphi$ is on b or $Tp\Vdash\psi$ is on b. Since φ, ψ are shorter than $\varphi \vee \psi$, by inductive hypothesis either φ is forced in $\mathcal{K}_b$ or ψ is forced in $\mathcal{K}_b$. By the definition of forcing, $\varphi \vee \psi$ is forced in $\mathcal{K}_b$.

2. Suppose that $Fp\Vdash\varphi \vee \psi$ is on b. Then b contains both $Fp\Vdash\varphi$ and $Fp\Vdash\psi$. Since φ, ψ are shorter than $\varphi \vee \psi$, by inductive hypothesis, p does not force φ in $\mathcal{K}_b$ and p does not force ψ in $\mathcal{K}_b$. By the definition of forcing, p does not force $\varphi \vee \psi$ in $\mathcal{K}_b$.

3. Suppose $Tp\Vdash\varphi \wedge \psi$ is on b. Then b contains $Tp\Vdash\varphi$ and $Tp\Vdash\psi$. Since φ, ψ are shorter than $\varphi \wedge \psi$, the inductive hypothesis says that p forces φ in $\mathcal{K}_b$ and p forces ψ in $\mathcal{K}_b$. By the definition of forcing, p forces $\varphi \wedge \psi$

4. Suppose that $Fp\Vdash\varphi \wedge \psi$ is on b. Then b contains either $Fp\Vdash\varphi$ or $Fp\Vdash\psi$. Since φ, ψ are shorter than $\varphi \wedge \psi$, the inductive hypothesis says that p does not force φ in $\mathcal{F}_b$ or p does not force ψ in $\mathcal{F}_b$. By the definition of forcing, p does not force $\varphi \wedge \psi$ in $\mathcal{F}_b$.

5. Suppose that $Tp\Vdash(\exists x)\varphi(x)$ is on b. Then there is a constant c such that b contains $Tp\Vdash\varphi(c)$. Since $\varphi(c)$ is shorter than $(\exists x)\varphi(x)$, the inductive hypothesis says that p forces $\varphi(c)$ in $\mathcal{F}_b$. The definition of forcing says that since c is in C(p), p forces $(\exists x)\varphi(x)$ in $\mathcal{F}_b$.

6. Suppose that $Fp\Vdash(\exists x)\varphi(x)$ is on b and c is a constant occurring in C(p). Then b contains $Fp\Vdash\varphi(c)$. Since $\varphi(c)$ is shorter than $(\exists x)\varphi(x)$, the induction hypothesis says that p does not force $\varphi(c)$ in $\mathcal{F}_b$. By the definition of forcing, p does not force $(\exists x)\varphi(x)$ in $\mathcal{F}_b$.

7. Suppose that $Tp\Vdash \varphi \rightarrow \psi$ is on branch b. Then for all $p' \geq p$, b contains either $Fp'\Vdash\varphi$ or $Tp'\Vdash\psi$. Since φ, ψ are shorter than $\varphi \rightarrow \psi$, the induction hypothesis says that p' does not force φ in $\mathcal{F}_b$ or p' forces ψ in $\mathcal{F}_b$ Since this is true for all $p' \geq p$ in P, the definition of forcing says that p forces $\varphi \rightarrow \psi$ in $\mathcal{F}_b$.

8. Suppose that $Tp\Vdash\neg\varphi$ occurs on b, and $p' \geq p$ occurs on b, then $Fp'\Vdash\varphi$ is on b. Since φ is shorter than $\neg\varphi$, p' does not force φ in $\mathcal{F}_b$. Since this is true for all $p' \geq p$ in P, the definition of forcing shows that p forces $\neg\varphi$ in $\mathcal{F}_b$.

9. Suppose $Tp\Vdash(\forall x)\varphi(x)$ occurs on b and $p' \geq p$ occurs on b and c is any constant in C(p'), then $Tp'\Vdash\varphi(c)$ occurs on b. By inductive assumption since $\varphi(c)$ is shorter than $(\forall x)\varphi(x)$, p' forces $\varphi(c)$ in $\mathcal{F}_b$. Since this is true for all constants c in C(p') and for all $p' \geq p$, p forces $(\forall x)\varphi(x)$ in $\mathcal{F}_b$.

10. Suppose $Fp\Vdash\varphi \rightarrow \psi$ is on b. For a $p' \geq p$, $Tp'\Vdash\varphi$ and $Fp'\Vdash\psi$ are on b. Since φ, ψ are shorter than $\varphi \rightarrow \psi$, by inductive hypothesis p' forces φ in $\mathcal{F}_b$ and p' does not force ψ in $\mathcal{F}_b$. The definition of forcing says p does not force $\varphi \rightarrow \psi$ in $\mathcal{F}_b$.

11. Suppose $Fp\Vdash\neg\varphi$ is on b. Then for a $p' \geq p$, $Tp'\Vdash\varphi$ is on b. Since φ is shorter than $\neg\varphi$, by inductive assumption p' forces φ in $\mathcal{F}_b$. The definition of forcing in $\mathcal{F}_b$ then says p does not force $\neg\varphi$ in $\mathcal{F}_b$.

12. Suppose $Fp\Vdash(\forall x)\varphi(x)$ occurs on b. Then there is a $p' \geq p$ and a constant c such that $Fp'\Vdash\varphi(c)$ is on b. Then since $\varphi(c)$ is shorter than $(\forall x)\varphi(x)$, by inductive assumption p' does

not force $\varphi(c)$ in $\mathcal{K}_b$. So the definition of forcing says p does not force $(\forall x)\varphi(x)$ in $\mathcal{K}_b$. □

COMPLETENESS THEOREM. If φ is forced in all frames, then φ has a tableaux proof.

PROOF. If the systematic tableaux with apex $F0\Vdash\varphi$ is not a proof, then by König's lemma there is an open branch b. By the theorem above 0 does not force φ in $\mathcal{K}_b$. Therefore φ does not hold in all frames. □

INTUITIONISTIC PROPOSITIONAL LOGIC DECISION METHOD

A statement φ of propositional logic has occurrences of only a finite number n of propositional letters. In classical logic such a φ is valid if and only if the 2^n truth assignments to these n propositional letters all extend to truth valuations in which φ is true. We can construct these 2^n valuations of φ and display them as the truth table of φ. Whether or not φ is valid is decidable by whether or not the last column of the truth table consists entirely of T's. For another proof of the decidability of the validity problem for classical propositional logic, to decide the validity of φ as in Smullyan [1968] construct a classical complete systematic tableaux with apex $T\neg\varphi$ by decoding all non–atomic entries systematically and stopping each branch when all non–atomic entries have been decoded. Either the resulting complete systematic tableaux is a classical tableaux proof of φ, or some branch is open. In the latter case, assigning α true for each atomic α with $T\alpha$ on that branch makes φ false. So φ is true in all truth valuations if and only if every branch of the complete systematic tableaux is closed. Is there an analogue decision method for validity for intuitionistic propositional logic? The key is given by the

FINITE MODEL PROPERTY. A statement is forced in all frames if and only if forced in all finite frames.

A somewhat tableaux oriented proof of decidability of the validity problem for intuitionistic propositional calculus goes as follows. Suppose we wish to determine the validity of φ, namely whether or not φ is forced in all frames. Construct a complete systematic tableaux starting $F0\Vdash\varphi$ as in a previous section. Either this complete systematic tableaux is a proof of φ, that is, all branches are closed, or there is an open branch whose occurring $Tp\Vdash\alpha$ with α atomic describe a (possibly infinite) frame in which φ is not forced. The finite model property says then there is in this case a finite frame in which φ is not forced. So if we effectively enumerate all tableaux proofs and simultaneously enumerate all finite frames and for each such frame determine whether or not φ is forced, by the above either a proof of φ will be found, or a finite frame in which φ is not forced will be found. In the first case φ is valid, in the second not, so this is a decision procedure for validity in intuitionistic propositional calculus. This method does not give a direct way of stopping the systematic tableaux procedure at a fixed finite stage of development, saying at that point that either the complete systematic tableaux has all branches closed and φ is valid, or is not forced in the frame associated with a remaining open branch as above. The finite model property is a direct consequence of the

FILTRATION LEMMA. Let $\mathcal{F}$ be a frame for propositional logic. Let X be a set of formulas containing with any formula all its subformulas. For p in P, define

$$[p] = \{q \in P \mid (\forall \varphi \in X)(p \text{ forces } \varphi \leftrightarrow q \text{ forces } \varphi)\}.$$

Let P_X be the set of all such [p] for p in P. Partially order P_X by $[q] \leq [p]$ if every formula of X forced by q is forced by p. (Due to the definition of [p], [q], this is the same as the requirement that every formula of X forced by some r in [q] is also forced by some s in [p]). Define a frame with P_X as partially ordered set and $A([p]) = A(p) \cap X$. Then for all φ in X, [p] forces φ if and only if every r in [p], r forces φ.

PROOF. We proceed by induction on the length of formulas.

BASE STEP. $A([q]) = A(q) \cap X$ says if φ is an atomic formula in X, then φ is forced by [q] if and only if φ is in $A(q) \cap X$, or if and only if φ is forced by q.

INDUCTION STEP. Suppose φ, ψ are in X. Suppose, for an induction hypothesis, that for all q, q forces φ if and only if [q] forces φ, and q forces φ if and only if [q] forces ψ

1) Implication. Suppose [p] forces $\varphi \to \psi$. We must show that p forces $\varphi \to \psi$. If $q \geq p$ and q forces φ, by induction hypothesis [q] forces φ, so by assumption [q] forces ψ, and by induction hypothesis q forces ψ. So p forces $\varphi \to \psi$. Conversely, suppose every r in [p] forces $\varphi \to \psi$ and $\varphi \to \psi$ is in X. We must prove that [p] forces $\varphi \to \psi$, that is, if $[q] \geq [p]$ and [q] forces φ, then [q] forces ψ. But $[q] \geq [p]$ and $\varphi \to \psi$ in X means that for every r in [p] there is an s in [q] forcing $\varphi \to \psi$. By induction hypothesis, [q] forces φ implies s forces φ. By definition of forcing s forces $\varphi \to \psi$ and s forces φ implies s forces ψ. By induction hypothesis, [q] forces ψ.

2) Negation is similar or can be reduced to the implication $\varphi \to F$.

3) Conjunction. p forces $\varphi \wedge \psi$ if and only if p forces φ and p forces ψ if and only if by induction hypothesis [p] forces φ and [p] forces ψ if and only if [p] forces $\varphi \wedge \psi$.

4) Disjunction is similar to conjunction. □

Why does it follow that the intuitionistic propositional calculus has the finite model property? A statement φ has only finitely many subformulas, say n of them. Thus there are only 2^n sets of subformulas, and each node in the filtered frame corresponds to one of them. So the frame is finite. However, we would hope for a better algorithm based on the tableaux development itself. Is there a finite stage in the development of a tableaux for intuitionistic propositional logic where we can safely stop adding new entries and conclude that all current open branches will remain open? This is what we require.

The filtration lemma gives us a hint as to how such a stage might be recognized. If a finite branch b is open, and will always be extendible by an open branch b', it gives us an initial segment of the Kripke frame $\mathcal{K}_{b'}$ associated with b', which has a filtration $\mathcal{F}$. Suppose we kept track of the emerging nodes [p] of the (possible) filtered frame for each branch, as our tableaux is being developed. Since the number of nodes in a filtered frame is bounded by the cardinality of the power set of the set of signed subformulas of the apex formula, there is a finite stage at which all such nodes in the filtered model will have turned up. If we can recognise such a stage we can stop. For the algorithm to work, we must keep track of all periodicities or repetitions of signed statements at each node on each path, to be sure that nothing new can be created. We now give the details.

To simplify the formulation of the algorithm we will modify the tableaux rules slightly, and make the development completely systematic.

DEFINITION. We define the MCST, the modified complete systematic tableau for intuitionistic propositional logic as follows.

1. Start the tableaux with apex $F0\Vdash\alpha$ as in the standard tableaux, where α is the proposition to be tested for validity.
2. Only attack the entries that <u>require attention</u> in the sense to be made precise below.
3. When permitted by the main algorithm, given below, use the following <u>canonical</u> <u>rules</u> of attack.

(i) If the entry being attacked is one of $Tp\Vdash\varphi\wedge\psi$, $Fp\Vdash\varphi\wedge\psi$, $Tp\Vdash\varphi\vee\psi$, $Fp\Vdash\varphi\vee\psi$ then append to the end of each open branch b through the entry the same signed subformulas already given in the standard tableau rules, e.g., for $Tp\Vdash\varphi\wedge\psi$, append

$$\begin{array}{c} Tp\Vdash\varphi \\ | \\ Fp\Vdash\psi. \end{array}$$

When done, declare the original entry <u>used</u>.

(ii) If the entry being attacked is $Tp\Vdash\varphi\rightarrow\psi$ then for each open branch b through the entry, and for each node q, with $q\geq p$, occurring on b, append

$$Fq \Vdash \varphi \qquad Tq \Vdash \psi$$

to the end of b and declare the original entry to be <u>locally used</u>.

(iii) Declare entries of the form $Fp \Vdash \varphi \to \psi$ to be <u>bad</u> entries.

Their use is defined in two stages:

<u>Stage I</u> For each open branch b through the entry, let p' be the least binary sequence strictly greater then p and incomparable with every q on b not $\leq$ p. Then append

$$\begin{array}{c} Tp' \Vdash \varphi \\ | \\ Fp' \Vdash \psi \end{array}$$

to the end of b.

<u>Stage II</u> For each open branch b through the entry being attacked and for each entry on b of the form $Tr \Vdash \xi \to \zeta$ which has already been declared <u>locally used</u> at some prior stage, with $r < p'$, append

$$Fp' \Vdash \xi \qquad Tp' \Vdash \zeta$$

to the end of b. This is called <u>re-using</u> $Tr \Vdash \xi \to \zeta$ on b for the sake of $Fp \Vdash \varphi \to \psi$.

This completes the definition of the canonical rules. We omit discussion of $p \Vdash \neg\varphi$ by rewriting $\neg\varphi$ as $\varphi \to F$. We discuss what to do when F is forced below.

We now proceed with the <u>main algorithm</u> which determines:

- which entries require attention,
- when the algorithm terminates,
- a minimization step, reflecting the monotonicity of forcing.

First a few definitions, conventions and ancillary rules.

DEFINITION. Recall that any entry of the form $Fp \Vdash \varphi \to \psi$ is called <u>bad</u>. All other kinds of entry are called <u>good</u>. Let $B \in \{T,F\}$. If $Bp \Vdash \varphi$ is an entry on a branch b, we say $B\varphi$ is a (signed) p–formula on b. Now define

$A_{p,b} = \qquad \{\varphi \mid Tr \Vdash \varphi$ is on b for some $r \leq p\} \cup \{F\varphi \mid Fp \Vdash \varphi$ is on b$\}$, i.e., all the signed p–formulas which are on b, or are implicitly there by monotonicity.

Define, for a partially developed tableaux $\mathfrak{T}$, the least entry requiring attention to be the leftmost, topmost entry not closed off (as defined below) on $\mathfrak{T}$ that is:

– good,
– nonatomic,
– unused (locally or globally),

if such an entry exists. Otherwise, it is the least unused entry.

A branch b is closed if, for some atomic statement φ and any $q \leq p$, $Tq \Vdash \varphi$ and $Fp \Vdash \varphi$ both occur on b, in which case it is marked with an X to indicate contradiction at the end, and never developed further. An entry ν is declared closed off if every branch through ν is closed. Every time an atomic entry is written down on a branch b, b should be checked for closure.

MINIMIZATION RULE. When attacking an entry, NEVER write down a new entry $Tp \Vdash \varphi$ on a branch b if for some $r \leq p$ $Tr \Vdash \varphi$ already occurs on b. (Observe than this does not affect our ability to detect closure). It is essential to respect the following rule when carrying out this "minimizing omission": All branchings must be preserved. For example, if attacking an entry would result in

$$Tp \Vdash \varphi \qquad Fp \Vdash \psi$$

being appended to a branch, and $Tp \Vdash \varphi$ is to be omitted because for some $r \leq p$, $Tr \Vdash \varphi$ has already shown up on b, we must write

$$\qquad\qquad Fp \Vdash \psi$$

and make sure the branching is preserved.

If we develop some signed entry for $p \Vdash \varphi \to F$ according to the rules, we never write down the entry $Fq \Vdash F$ for any q, and we declare a branch b closed the moment $Tq \Vdash F$ is written down.

THE MAIN ALGORITHM. Proceed by attacking the least entry requiring attention via the canonical rules subject to one restriction which we now describe. It will be convenient to thing of the attack on a given entry ν as being carried out in steps, one for each open branch b passing through ν. We then stipulate that no entry is to be appended to a branch b when attacking a bad entry $\nu = Fp \Vdash \varphi \to \psi$ if ν is b–finished. We define such a bad entry ν to be b–finished, or finished for branch b if:

1. All good entries on b have been used,

2. There is an $r < p$ on b such that every unused r–formula on b has been declared b–finished, and

3. $A_{r,b} = A_{p,b}$.

In such a case, we say ν is b–finished for the sake of r. We declare a bad entry ν to be finished if it is used or b–finished for every branch b through ν.

The algorithm terminates (if ever) when every entry on the tableaux is either used or finished.

We give three lemmas which say that the algorithm just described achieves its aims.

LEMMA 1. The MCST algorithm always terminates.

PROOF. Suppose not. Then there is, by König's Lemma, an infinite branch b. An inspection of the rules shows that no node p can appear infinitely often on b. An attack on a good formula always results in a reduction in the depth (complexity) of a formula. Attacking a bad formula does not, since, in general, it forces the re–use of prior formulas of possibly greater complexity. But each use of a bad formula generates strictly greater nodes. Thus, at any stage in the development of a branch b, for a given node p, once a p–entry has been written down, new p–formulas can only be introduced by the use of good formulas on b and the good formulas they produce, necessarily of lower complexity. Thus, after finitely many steps, only bad formulas remain, and then, there will never again be an occasion to produce p–formulas on that branch.

Thus, if b is infinite, there are infinitely many distinct nodes on b. These nodes determine a subtree, $\mathbf{P}_b$, of the full binary tree (namely the underlying partial order of the Kripke frame associated with b) so, again by König's lemma, there is an infinite strictly increasing chain $p_0 < p_1 < \dots < p_n < \dots$ on b. For each node p_i in this chain, there must have been at least one bad p_i–formula on b that was not declared finished on the sub–branch b_i of b that existed when it was attacked, otherwise there would have been no reason to introduce p_{i+1} on b. Thus, the set A_{p_i,b_i} was found to be distinct from all A_{p_j,b_i} for $p_j < p_i$ on b_i. As remarked above, when a bad p_i–formula is attacked (under MCST rules) there will never again be an occasion to introduce new p_i–formulas on any branch extending b_i , hence $A_{p_i,b_i} = A_{p_i,b}$, i.e., p_i is saturated on b (all p_i–formulas which will ever show up on b have already shown up). Arguing the same way for every i, $\{A_{p_i,b} \mid i \in \omega\}$ must be an infinite sequence of distinct sets of signed subformulas of the original formula α at the apex. But this is impossible, since the set of subformulas of α is finite. (In fact, if the set S_α of signed subformulas of α has cardinality n, the sequence $A_{p_i,b}$ must repeat for some $i, j \leq 2^n$). Thus the algorithm terminates. □

LEMMA 2. The MCST algorithm is sound for intuitionistic propositional logic.

PROOF. If every branch of the MCST for $F0\Vdash\alpha$ is closed then α is intuitionistically valid". The reason a branch closes in the MCST is the same as for the ordinary (un–modified) tableaux, already proven sound. □

LEMMA 3. The MCST algorithm is complete for intuitionistic propositional calculus.

PROOF. We must argue that if a branch b is terminated and left open in the MCST procedure because every unused formula on b was declared b–finished, then if we extended the branch by the standard tableaux rules, we would not close off b, i.e. some b' extending b would still be open. Since we already know that standard tableaux deduction is complete, it would follow that there is a frame agreeing with that branch, thus establishing completeness of the MCST as a proof procedure.

Let $\nu = Fp\Vdash\varphi\to\psi$ be the first entry on b that was declared finished. (If there were none, b is open by standard tableau rules and we would be done). If at the time this occurred, the sub–branch b' of b had been developed, then for some $r < p$ on b',

$$A_{r,b'} = A_{p,b'}.$$

By the remarks above, neither of these sets can increase, since all good formulas on b' had to be used prior to consideration of a bad one. Thus $A_{r,b} = A_{p,b}$. By our minimization rule, this also means that no true–signed formulas $Tr'\Vdash\theta$ occurs on b for $r < r' \leq p$. We know that $Fr\Vdash\varphi\to\psi$ occurred earlier on b and was attacked and used. Suppose we ignore the fact that $Fp\Vdash\varphi\to\psi$ is finished and develop it according to standard tableaux rules. (We don't touch the other bad entries on b which have been declared finished, because each of them would result, if used, in nodes incomparable to those resulting from any of the others. Since a contradiction involves comparable nodes, it is easy to see that they will not interfere with one another, and that any contradiction generated would be due to the development of <u>one</u> of them alone). Now consider the branches generated by using ν. One of them, say d, corresponds to the path b itself in the sense that every new entry made on d corresponds to an entry made on b for the sake of $Fr\Vdash\varphi\to\psi$. This includes any true–implications that were re–used when $\nu = Fp\Vdash\varphi\to\psi$ was attacked. (The reader can show that any such locally used $Ts\Vdash\xi\to\zeta$ on b must satisfy $s \leq r$ and must have already shown up when $Fr\Vdash\varphi\to\psi$ was attacked). Thus, either d ends without contradiction due to the use of only good entries, or one of the unused bad formulas on b shows up again. It will either be a fresh copy of $F(\varphi\to\psi)$, in which case periodicity is guaranteed, or it is one of the others. Continuing this argument for the others, we can show that periodicity will eventually occur later, in the same way. □

EXERCISE Supply the missing details in the proof of the completeness of the modified tableaux rules.

REMARK. For polynomial space completeness of the decision problem for intuitionistic propositional calculus, see Statman [1979].

PRENEX FORMULAS

A prenex formula is one of the form

$$Q_1 \ldots Q_n \varphi$$

where φ is quantifier–free, i.e. all quantifiers are in front. In classical predicate logic, every formula is logically equivalent to a prenex formula, and this is the basis of the introduction of function symbols and Herbrand terms. This avenue is not available in intuitionistic logic because not every formula of predicate logic is logically equivalent to a prenex formula. The reason is this: There is an algorithm which, applied to a prenex statement φ, determines in a finite number of steps whether or not φ is valid. But there is no algorithm which, applied to arbitrary statements of predicate logic, determines whether or not they are intuitionistically valid. This assertion can be reduced to the corresponding problem for classical logic by embedding classical logic in intuitionistic logic using Gödel's translation, which maps classically valid statements to intuitionistically valid statements using $\neg\neg$.

EXERCISE SET 4.

1. (Gödel's Translation). To each predicate logic formula φ make correspond another predicate logic formula φ^o defined by the following inductive clauses.

0) F^o is F
1) φ^o is $\neg\neg\varphi$ for for atomic φ other than F.
2) $(\varphi \wedge \psi)^o$ is $\varphi^o \wedge \psi^o$
3) $(\varphi \vee \psi)^o$ is $\neg(\neg\varphi^o \wedge \neg\psi^o)$
4) $(\varphi \rightarrow \psi)^o$ is $\varphi^o \rightarrow \psi^o$
5) $((\forall x)\varphi)^o$is $(\forall x)\varphi^o$
6) $((\exists x)\varphi)^o$ is $\neg(\forall x)\neg\varphi^o$

Using frames prove by induction on length of formulas that φ is classically valid if and only if φ^o is intuitionistically valid.

2. Use the disjunction and existence properties for intuitionistic predicate logic to show that the decision problem for prenex formulas in intuitionistic logic can be reduced to the decision problem for intuitionistic propositional logic given above.

CODING FRAMES INTO CLASSICAL MODELS

There is a translation which makes frames the study of the models of a related classical predicate logic theory K. We defined a frame $\mathcal{F}$ as a partially ordered set P with partial order $\leq$ and associating with each p in P a set C(p) and a set A(p) of the atomic statements $R(c_1,...,c_n)$ with $c_1,..., c_n$ in C(p) such that A(p), C(p) are monotone increasing in p. Make frames into classical models by making the partially ordered set P into a "first class citizen" by introducing the following predicates.

1. A unary predicate P(x), meaning x is in the partially ordered set.
2. A binary relation $\leq$ for the partial ordering.
3. A binary relation C(p, x), meaning x is in domain C(p).
4. An n+1–ary relation symbol R^* for each n–ary relation R, so $R^*(p, c_1,...,c_n)$ means that $R(c_1,...,c_n)$ is in A(p).

We want the domain of a model of the related first order theory K to be the disjoint union of P and all the C(p).

$$(\forall x)(P(x) \vee (\exists p)C(p, x))$$
$$(\forall x)(\neg(P(x) \wedge (\exists p)C(p, x)).$$

We want the first argument of R^* to be from P and the rest of the arguments to be from C(p).

$$(\forall p)(\forall x_1)...(\forall x_n)(R^*(p, x_1,...,x_n) \rightarrow P(p) \wedge C(p, x_1) \wedge ... \wedge C(p, x_n))$$

We want $\leq$ to partially order P and have no contributions from the C(p).

$$(\forall p)(\forall q)(p \leq q \rightarrow (P(p) \wedge P(q))$$
$$(\forall p)(\forall q)(p \leq q \wedge q \leq p \rightarrow p = q)$$

We want the frame requirement that C(p) is monotone in p.

$$C(p, x) \wedge p \leq p' \rightarrow C(p', x)$$

We want the frame requirement that A(p) is monotone in p, but expressed in terms of the first argument of R^*.

$$(\forall p)(\forall p')(\forall x_1)...(\forall x_n)(P(p) \wedge P(p') \wedge p \leq p' \wedge C(p, x_1) \wedge \quad \wedge C(p, x_n)$$
$$\rightarrow R^*(p, x_1,...,x_n) \rightarrow R^*(p', x_1,...,x_n)).$$

For a fixed intuitionist predicate logic let K be the set of corresponding axioms in classical logic as above, using the newly introduced relation symbols in place of the old ones of the intuitionistic logic. Then every frame uniquely translates to a classical model M of K and every classical model M of K translates uniquely to a frame.

Here is a translation of forcing statements about a frame into the language of K. (See Van Dalen's review article [1986].)

0) $(p \Vdash F)^t$ is F.
1) $(p \Vdash R(x_1,\dots,x_n))^t$ is $R^*(p,x_1,\dots,x_n)$
2) $(p \Vdash (A \to B))^t$ is $(\forall q)(q \geq p \wedge (q \Vdash A)^t \to (q \Vdash B)^t)$
3) $(p \Vdash \neg A)^t$ is $(\forall q)(q \geq p \to \neg(q \Vdash A)^t)$
4) $(p \Vdash (\forall x)P(x))^t$ is $(\forall q)(\forall x)((q \geq p) \wedge C(q, x) \to (q \Vdash P(x)^t)$
5) $(p \Vdash (\exists x)P(x))^t$ is $(\exists x)(C(p, x) \wedge (p \Vdash P(x))^t)$
6) $(p \Vdash A \wedge B)^t$ is $(p \Vdash A)^t \wedge (p \Vdash B)^t$
7) $(p \Vdash A \vee B)^t$ is $(p \Vdash A)^t \vee (p \Vdash B)^t$

THEOREM. Let M be the classical model of K associated with a frame. Then for any p in P, p forces φ in the frame if and only if $(p \Vdash \varphi)^t$ is true in M.

PROOF. By induction on length of formulas. The translation is chosen to make this correct.
□

REMARK. A statement φ is intuitionistically valid if and only if in every classical model with K valid, $(\forall p)(p \Vdash \varphi)^t$ is valid. This has as a consequence that any complete proof procedure for classical predicate logic yields a complete proof procedure for intuitionistic logic. For example, φ is intuitionistically valid if and only if there is a classical logic tableaux deduction of $(\forall p)(p \Vdash \varphi)^t$ from K. For another example, suppose resolution plus unification is the proof procedure of choice for classical logic. Then φ is intuitionistically valid if and only if there is a resolution–unification proof of $(\forall p)(p \Vdash \varphi)^t$ from K. With small changes to improve efficiency, this is a simple way of implementing intuitionistic logic deduction using software implementing classical logic deduction. A similar observation applies to many other modal logics.

Let T be a (say) countably infinite set of statements in classical logic., let ψ be one statement. We know by countable compactness of classical predicate logic that ψ is true in all models of T if and only if there is a finite $T' \subseteq T$ such that ψ is true in all models of T'. With the translation given above, this now applies to intuitionistic logic.

COROLLARY (Countable compactness). Let T be a countable set of statements in intuitionistic predicate logic. Then ψ is forced in all frames in which T is forced iff there is a finite subset T' of T such that ψ is forced in all frames in which T' is forced.

This says ψ is a semantic consequence of T iff ψ is a semantic consequence of a finite subset of T.

Of course, countable compactness can also be proved easily by tableaux. We omit the set–theoretic discussion necessary to extend compactness to uncountable sets of propositions.

[1]supported by NSF grant DMS–89–02797 and ARO contract DAAG29–85–C–0018

BIBLIOGRAPHY

Starred entries (*) contain extensive bibliographies.

Part I: General References

M. J. Beeson(*) [1985], *Foundations of Constructive Mathematics*, Springer–Verlag, Berlin.
An encyclopedic reference summarizing the metamathematics of constructive mathematics. The exposition is based on partial applicative structures as models for realizability. This treatise explains many of the constructive formal systems of use in computer science such as Martin–Löf's and Feferman's, and also their recursive realizability and forcing interpretations. Eschews the Kripke frame approach. There are inaccuracies in a few definitions and proofs, but this is a very useful book anyway due to its unified point of view on material that is otherwise quite scattered.

M. Dummett [1977], *Elements of Intuitionism*, Clarendon Press, Oxford.
Thorough statement of the philosophical principles behind intuitionism, as well as a detailed introductory exposition of the proof theory and Beth and Kripke semantics.

A. G. Dragalin(*) [1987], *Mathematical Intuitionism: Introduction to Proof Theory*, Translations of Mathematical Monographs 67, AMS, Providence, R. I.
Assumes no background and gives some unity to Kripke frames and Kleene's realizability semantics.

M. Fitting(*) [1983], *Proof Methods for Modal and Intuitionistic Logics*, D. Reidel, Dordrecht.
Studies tableaux proof procedures via consistency properties in the Smullyan tradition for over a dozen modal and intuitionistic logics, as well as decision methods and interpolation. This book is the basis for much recent work in automated deduction for nonclassical logics.

D. G. Gabbay [1981], *Semantical Investigations in Heyting's Intuitionistic Logic*, D. Reidel, Dordrecht.
An in–depth study of Kripke frame semantics, including decidability and undecidability of various theories and a discussion of recursive and r.e. presentations of frames. This is a good beginning for the still underdeveloped model theory of intuitionistic systems.

J–Y. Girard [1987], *Proof Theory and Logical Complexity*, vol. I, Studies in Proof Theory, Bibliopolis, Naples.
An introduction emphasizing the ordinary mathematical structure of proof theory more than any other text currently available.

A. Heyting [1956], *Intuitionism: an Introduction*, North–Holland, Amsterdam.
The standard exposition of 1930's intuitionism.

S. C. Kleene [1952], *Introduction to Metamathematics*, North–Holland (1971 edition), Amsterdam.
This classic text still has the best exposition of the relation between classical and intuitionist logic.

A. Nerode(*) [1986], "Applied logic", in *The Merging of Disciplines: New Directions in Pure, Applied, and Computational Mathematics*, (R. E. Ewing, K. I. Gross, C. F. Martin, eds.), Springer–Verlag, Berlin, 127–163.
A brief guide to some applications of logic to computer science as of 1986.

A. Nerode and J. B. Remmel(*) [1985], "A survey of r.e. substructures", in *Recursion Theory* (A. Nerode and R. Shore, eds.), Proc. Symposia in Pure Math 42, Amer. Math. Soc., Providence, RI, 323–373.
This article gives an introduction to the methods of recursive algebra, a more completely developed classical analogue of intuitionistic algebra. Recursive algebra is linked to intuitionistic algebra through recursive realizability.

D. Prawitz [1965], *Natural Deduction*, Almqvist and Wiskell, Stockholm.
This exposition of natural deduction eventually made the isomorphism of natural deduction calculi with suitable typed lambda calculi transparent.

R. M. Smullyan [1968], *First–Order Logic*, Springer–Verlag, Berlin.
The first part of this book is a very usable textbook on classical logic tableaux.

A. S. Troelstra(*) [1973], *Metamathematical Investigation of Intuitionistic Arithmetic and Analysis*, Mathematical Lecture Notes 344, Springer–Verlag, Berlin.
The encyclopedic reference to the subject as developed before 1973. Includes a book–length chapter on Kripke frames, by Smorynski, and chapters on almost every other aspect of the subject known at the time. This is a cumbersome but necessary reference.

A. S. Troelstra and D. van Dalen(*) [1988], *Constructivism in Mathematics*, vol. I, II, North–Holland.
A very comprehensive and generally lucid treatment of intuitionistic logic and mathematics. Much material here is original or otherwise available only in non–uniform notation in hard–to–locate papers of many authors.

D. van Dalen [1983], *Logic and Structure*, second ed., Springer–Verlag, Berlin.
An undergraduate logic text with an excellent chapter on intuitionistic logic and frames.

D. van Dalen(*) [1986], "Intuitionistic logic", in *The Handbook of Philosophical Logic*, vol. III, 225–339, D. Reidel, Dordrecht.
The best succinct introduction to the mathematics of intuitionistic logics yet written.

Part II: Other References

E. Bishop [1967], *Foundations of Constructive Analysis*, McGraw–Hill, New York.
The 1985 edition is much expanded with D. Bridges as co–author.

H. de Swart [1976], "Another intuitionistic completeness proof", J. Symb.Logic 41, 44–662.

M. P. Fourman and D. S. Scott [1977], "Sheaves and logic", in *Applications of Sheaves*, (Fourman, Mulvey and Scott, eds.), Mathematical Lecture Notes 753, Springer–Verlag, Berlin, 302–401.
This paper is very useful for understanding the relation between higher order intuitionistic logic and category theory.

J–Y. Girard [1971], "Une extension de l'interprétation de Gödel à l'analyse, et son application à l'élimination de coupures dans l'analyse et la théorie des types", in *Proceedings of the Second Scandinavian Logic Symposium*, (J. E. Fenstad, ed.), North–Holland, Amsterdam.

J–Y. Girard [1972], "Interprétation fonctionelle et élimination de coupures de l'arithmétique d'ordre supérieur," Thèse de Doctorat d'état, Université Paris VII.

J–Y. Girard, Y. Lafont, and P. Taylor [1989], *Proofs and Types*, Cambridge tracts in theoretical computer science 7, Cambridge University Press, Cambridge, England.
The first exposition in understandable textbook form of Girard's system F as a basis for polymorphic λ–calculus. This book also contains some linear logic.

R. J. Grayson [1983], "Forcing in intuitionistic systems without power set", J. Symb. Logic, 48, 670–682.

J. E. Hopcroft and J. D. Ullman [1979], *Introduction to Automata Theory, Languages and Computation*, Addison–Wesley, Reading, Mass.

W. A. Howard [1980], "The Formulae–as–types notion of construction", in *To H. B. Curry: Essays in Combinatory Logic, Lambda Calculus and Formalism*, Academic Press, 479–490.

G. E. Hughes and M. J. Cresswell [1968], *An Introduction to Modal Logic*, Methuen, London.

S. C. Kleene and R. Vesley [1965], *The Foundations of Intuitionistic Mathematics*, North–Holland, Amsterdam.

G. Kreisel [1962], "On weak completeness of intuitionistic predicate logic", J. Symb. Logic 27, 139–158.

S. Kripke [1965], "Semantical analysis of intuitionistic logic I", in *Formal Systems and Recursive Functions*, (J. N. Crossley, M. Dummett, eds.) North–Holland, Amsterdam, 92–130.

J. Lambek and P. J. Scott [1986], *Introduction to Higher Order Categorical Logic*, Cambridge Studies in Advanced Mathematics 7, Cambridge University Press, Cambridge, England.

H. Läuchli [1970], "An abstract notion of realizability for which predicate calculus is complete," in *Intuitionism and Proof Theory*, (Kino, Myhill, Vesley, eds.), North–Holland, Amsterdam.

D. Leivant [1983], "Polymorphic type inference", in *Proc. 10th ACM Symposium on the Principles of Programming Languages*, ACM, New York, 88–98.

D. Leivant [1983a], "Reasoning about functional programs and complexity classes associated with type disciplines," in *24th Annual Symposium on Foundations of Computer Science*, 460–496.

H. Rasiowa and R. Sikorski [1963], *Mathematics of Metamathematics*, Monographie Matematyczne, t. 41, Warsawa.

C. Smorynski [1973], "Applications of Kripke models", in Troelstra [1973], see part I of the bibliography.

R. Statman [1973], "Intuitionistic propositional logic is polynomial space complete", Theoretical Computer Science 9, 67–72.

S. Stenlund [1972], *Combinators, λ–terms and Proof Theory*, D. Reidel, Dordrecht.

P. B. Thistlewaite, M. A. McRobbie, R. K. Meyer [1988], *Automated Theorem Proving in Non–Classical Logics*, Wiley, N.Y.

W. Veldman [1976], "An intuitionistic completeness theorem for intuitionistic predicate logic" J. Symb. Logic 41, 159–166.

L. Wallen, [1990], *Automated Proof Search in Nonclassical Logics*, MIT Press, Cambridge, Mass.

Making Computers Safe for the World: An Introduction to Proofs of Programs Part I

Richard A. Platek
Odyssey Research Associates, Inc.
Cornell University

Contents

List of Figures

1 Preface

These lectures were given in June 1988 at a summer school in Logic and Computer Science organized by the Italian research organization CIME. The summer school took place at Montecatini Terme, a charming spa in Tuscany. I would like to thank the scientific director of the meeting, Piergiorgio Odifreddi, of the University of Turin, for his various kindnesses and the staff of CIME for their patience while I prepared these lectures for publication amidst a demanding schedule. My colleagues at Odyssey Research Associates are a constant source of inspiration and criticism.

The present lectures are only the first part of a longer work which develops the theory of program verification from its logical foundations. This part is divided into two sections; one discursive, polemical, the other technical. In the former, I describe the rather deplorable state of software correctness and the risks to life and property which result from increased dependence on such trash. As I mention there, while the general public appears to accept mental sloppiness as a part of life, it is shocking to trained mathematicians to see intellectual products containing known errors peddled in the marketplace. The shock only deepens when one discovers that the software industry contains an uncomfortable number of one's former C students who, having squeaked through undergraduate school, have subsequently landed jobs writing life-critical software. It is our contention that no matter what slick technology might be introduced to increase the quality of software there is "no royal road" [1] to the correctness of intellectual claims other than proofs.

In the technical section we develop, in a novel way, the basic results underlying program verification. We work within the context of programs written in a flowchart language. Our results follow rather easily from providing a semantics for these programs based on inductive definability. Such a semantics allows us to compare programs in this language with logic programs (i.e., Prolog). In the second Part of this paper, which will appear separately, we extend our approach to structured programming languages, like Ada, by mapping such programs into our flowchart language and we encapsulate and reuse code through subroutines.

[1] Although probably apocryphal, the story goes that the Hellenistic King Ptolemy asked Euclid for a way to learn geometry simpler than the deductive method (being the heir to the Egyptian civilization he would have perhaps preferred animated graphics). Euclid supposedly answered that there is no Royal Road to truth. It is obviously a story with strong appeal to mathematicians who might feel neglected by the powers that be.

2 Introduction

2.1 The Problem

We all make many mistakes and then correct some of them. Although the purpose of technology is to implement our intentions, it also amplifies our inattentions. Unfortunately, technology is a non-linear magnifier; little mistakes can lead to large consequences. This is especially true of digital technology such as computers and software. Engineering has evolved a general testing methodology based on the empirical tradition in science. Artifacts are placed in a number of prepared environments and their behavior is observed. Although the testing is sometimes called "exhaustive" only a small number of experiments are actually performed compared to the larger number (infinite?) of possible ones. One then extrapolates in order to predict behavior in the untested, intermediate cases. As an engineering discipline, software has inherited this traditional test and evaluation methodology; it even has a fancy name "quality assurance". But digital technology can't be comprehensively tested in this manner since its responses are not continuous functions of input and therefore the behavior in the cases not actually examined can not be predicted by extrapolation. This is well understood by people who specialize in testing and they have developed techniques which strive to give adequate "coverage" of all classes of possibilities. Nevertheless, testing is almost always inadequate. What are the consequences of inadequate assurance methodologies?

The Therac 25 is a medical instrument used in the treatment of cancer. Essentially, it is a linear accelerator designed to send X-rays and electron beams deep into human bodies. Its aim is to destroy tumors without damaging surrounding tissue or skin. The system is computer controlled in order to achieve the desired degree of precision. Recently (1985-1986), the machine failed in three separate instances killing two patients and severely injuring a third. The malfunction was caused by a software error not detected during the extensive testing required of medical equipment by government regulating and licensing agencies. The error, which increased dosage levels by a factor of 100, surfaced only when the human operator entered an unusual sequence of keystrokes.

Some more horror stories: an F-18 fighter airplane crashed due to a missing exception condition in its software; a Mercedes 500SE with a graceful stop, no-skid, brake computer left a 368-foot skid mark in an accident which killed the passenger; Cadillac recalled 57,000 cars with a headlights-out computer problem; an improper software upgrade disrupted telephone service in the Poughkeepsie, NY area for 21 hours; an 18.5 million dollar rocket was destroyed because the Atlas-Agena software was missing a hyphen; the U.S. Strategic Air Command was put on alert due to an adequate software response to a chip failure at NORAD.

The following recall notice was recently sent out by Volvo:

> Volvo has determined that a defect which relates to motor vehicle safety exists in Volvo installed cruise control systems of 1986 and 1987 Volvo automobiles.
>
> In laboratory tests, we have been able to induce a malfunction in the microprocessor of the cruise control unit. We have found that if the cruise control switch is left in the "on" position and the car's electrical system experiences a voltage drop, the cruise control may unexpectedly engage. We have also

> found that the application of the brake pedal and the movement of the switch to the "off" position cancels the malfunction. This cannot occur if the cruise control is in the "off" position.
>
> We know of no cases where this has happened in normal driving, but we do not want a malfunction of the cruise control to contribute to an accident. Accordingly, we will replace the microprocessor of your cruise control at no charge to you.

The really interesting question is, how did Volvo discover this error?

The stories go on and on: robots killing workers; credit systems damaging individual and corporate credit ratings; overbilling; underbilling; uncontrolled fluctuations in financial markets due to programmed trading; etc., etc.

The above disaster stories are instances of "nature" exploiting design errors (i.e., there was no conscious determination on anyone's part to produce the undesired effects). But unintentional misfeasance invites intentional malfeasance (in simple English: don't leave your car unlocked). Information and communication technology have been implicated in cases of fraud, theft of money and information, and unauthorized alteration of data to either further personal ends or for pure maliciousness. In many cases software "errors" made such acts possible. In some cases, the error was a true code or design flaw which either permitted entry to unauthorized users or allowed authorized users to upgrade their status to a systems administrator role giving them free reign over the system. In other cases, the error was a failure to design the system so that misuse by legitimate users would be minimized. For example, although commercial systems do separate users from systems administrators there is not enough attention given to principles like "limitations of privileges" which constrains users to just those usage scenarios necessary to perform their functions. It's very hard to steal money from an automatic teller machine with just three buttons. Banks don't trust customers but they systematically overtrust employees. In addition, most systems have inadequate auditing and monitoring facilities. Trying to reconstruct events from low-level audit information is extremely tedious. What is needed is higher-order auditing facilities which would automatically and concurrently generate relevant summaries of high impact activity and detect irregularities and intrusion.

In addition to the many documented cases of serious computer malfunctions there is also a large body of apocryphical tales widely accepted as true. Like all folktales, the undocumented stories are readily believed because they sound like they could be true and something within us espects them to be true. For example, many people believe that the recent automobile sudden acceleration problem is due to a subtle software timing error causing automobile companies to pay very close attention to their embedded computers as illustrated by the Volvo recall notice quoted above.

As a literary genre, such cautionary tales are not new. From the domestication of fire to bio-engineering, the introduction of new technology has been viewed as a dangerous undertaking. The tragic figure of Prometheus, punished for stealing fire from the Gods, continually reappears.

Although a significant level of safety is necessary for widespread acceptance of a new technology there is an inevitable time lag between the development of new capabilities and the safety engineering which aims to minimize the associated risks. In the interim, there is fear and uncertainty which undermine the will to proceed. Highly publicized software failures fuel an ever increasing concern over the wisdom of trusting computers in critical situations. This concern is constantly being expressed in debates over computer involvement in: defense; automated plants (e.g., chemical factories, nuclear power plants); aviation; space; health care; consumer products; financial services; etc.

2.2 Towards a Solution

A software safety industry is emerging in response to these concerns. Diverse techniques are being explored as aids to increase confidence in systems: mathematical modeling; simulation; concurrent run-time monitors which check for error states; N-version programming; etc.

Many of these techniques address concerns broader than safety. For example, the costs of software maintenance presently outweigh development costs. One reason is the lack of modularity at the code level and the absence of formal specifications which describe in a declarative manner the function of these modules. Increased formality, which we are promoting as a safety and assurance method, would also contribute to solving the maintenance problem. Indeed, the classical development/maintenance distinction is currently being eroded by the emergence of a more evolutionary concept of the software life cycle and to be successful, such a paradigm will have to make use of formal methods.

2.2.1 Formal Methods

One of the most intellectually exciting areas being pursued for safety and other reasons is the use of formal methods; that is, a heavy use of the techniques of mathematical logic in the specifying and verifying of software.

In many ways programs are like proofs; for example, the analogy

$$proof : theorem :: program : specification$$

is very suggestive – to show a theorem is correct one constructs a proof, to meet a specification one constructs a program. Unfortunately, such analogies are merely suggestive, the cultural realities are quite different. Proofs which don't prove their theorems have little value within the mathematical community and are promptly withdrawn. Software, on the other hand, which fails to meet its specifications usually represents a considerable financial investment and continues to circulate. An incorrect proof is no proof at all; an incorrect program is still a program, it does something and can be delivered. It comes as quite a shock to academic mathematicians when they discover this fact of life. It's as if a graduate student who failed to make any progress on his thesis were to demand some kind of degree corresponding to the time and effort expended. All the false leads and discarded proofs would be bound together and submitted as an "effort-" or "pseudo-thesis". Unfortunately, a great deal of fielded software is on the level of such "pseudo-theses".

The main difference between programming and proving is that, aside from a few exceptional circumstances, the mathematical community can determine with amazing unanimity whether a purported proof is correct or not. Indeed, that's what makes mathematics mathematics. No other discipline is as non-controversial. This century's impressive achievements in mathematical logic and the general acceptance of the axiomatic approach throughout mathematics has served to enhance this ability.

The software community, on the other hand, does not possess an analogous judgmental ability with respect to its products. Even very high level language programs are not self-documenting. As opposed to proofs, these programs resemble cryptological codes which take considerable intellectual effort to break. In particular, they lack the discursive connecting tissue which would enable the reader to reconstruct the writer's thinking. Good programming style recommends the extensive use of informal comments to carry the reader along but such informal comments remain merely suggestive; no methodology enforces their truth. Indeed, the comments themselves need not have any meaning.

What is necessary is the development of:

- mathematical models of information systems and their environments; such modeling is especially important in embedded systems, i.e., systems like medical equipment in which the software is only a part; modeling should make use of the full repertoire of mathematics such as probabilistic models of the environment

- formal mathematical semantics for programming languages; most programming languages have no formal semantics, meanings are defined by compilers. Although tests for compiler acceptance and certification are in use it is well-known that they accept compilers which deviate from the language manual

- semantically based formal specification and design languages; such design languages should span the full spectrum from very high level conceptual designs (sometimes called "systems architectures") to program language level specifications

- proof techniques which enable one to prove that levels of a design are consistent with one another and that the resulting program meets its specifications

- computer aided deduction systems which automate most of the tedium and act as a verifier's assistant.

There are several trends which contribute to the realization of this vision. In the programming language community trading power for safety has become a recognized principle. Early high-order programming languages provided all sorts of tricks as power tools. Languages were viewed as means for taking maximum advantage of hardware. Most of these tricks relied on the fact that everything internal to a machine is stored as bit strings or an array of bytes. These languages were essentially type free and any value (such as a Boolean) was acceptable as an argument by any operator. Even when objects were typed, implicit type conversion was assumed to take place when necessary in order that expressions like *true* + 5 be capable of evaluation. This freedom was promoted as a productivity enhancer and the style still has its supporters. More recent languages, on the other hand, such as Ada, follow a hard typing discipline; all type conversions must be explicit. For example, Ada has a class of types known as "derived types" which make a "new" copy of an old type distinct from its prototype. So one can declare

```
type APPLE is new INTEGER;
type ORANGE is new INTEGER;
X : APPLE := 5; --this initializes the variable X to 5
Y : ORANGE;
```

but the assignment "Y := X" will be illegal and detected at compile time since 5 APPLEs do not an ORANGE make even though the same "raw" INTEGER type underlies both APPLE and ORANGE. Such restrictions are felt as onerous to some programmers but they serve to prevent a large class of stupid errors and to make code more intelligible. The run-time overhead for such hard typing is negligible; it just makes parsers more difficult to write. The parsers must keep track of a great deal of semantical information. If the text is legal it runs just as the type free code runs. A semantical obstacle is put in the way of the programmer; he can not get to the machine until certain tests are passed. This is a kind of programming proving and thus a trend in the direction we are indicating.

Academically, specification language principles and design are beginning to be considered as important as programming language principles and design. The experience of mathematical logicians is useful here since such formal specification languages are declarative rather than imperative. The logic of programs has been a core subject in computer science since the pioneering papers of Floyd and Hoare. Automated deduction is also receiving increased academic attention. These and similar contributions are being brought together through entrepreneurial initiative in a manner which will have a decisive impact on the way information systems are constructed.

Just as contemporary programs must parse (i.e., be syntactically correct) before they can compile we believe that a more careful approach to software will require semantic correctness evidence (i.e., proofs) as part of the compilation process. To a remarkable degree computer science has achieved the automation of syntax; it is now turning its attention to the automation of semantics.

2.3 The World Tomorrow

Formal verification need not entirely occur before compilation. A "verifying" compiler could generate concurrent, run-time checking code for those compilation units which have not been proven. Such code checks that a given run is meeting its formal specification. Run-time checking is an overhead expense. On the parallel hardware architectures that are emerging, such run-time checking will use up some of the processors (as will various fault-tolerant checking which attempts to correct for true "acts of God" as opposed to human intellectual weakness). As proofs are furnished these units with their proofs can be recompiled to form a more efficient system (more efficient in that fewer processors need be committed to correctness overhead). In this manner the verification process extends throughout the software life cycle. Indeed, the very first version of a program might contain no proofs; the run-time checking being used extensively for testing and debugging. Proofs would be attempted when one gains confidence in the code. An "intelligent" system should be able to use failed proofs to suggest (or even automatically generate) correct code segments. Modularity of proofs should match modularity of design and implementation in order to control the need for reverification during system upgrade.

Similar evolutionary thinking is emerging in the area of simulation. In the beginning one simulates a complete system and its environment to exercise the conceptual design. As the system is built pieces of it replace their simulation. Such simulation continues even after the system is completed; ultimately one is running the actual system against a simulated environment. Such an evolutionary test bed is a powerful software engineering tool. Like verification it would require more computational power than that used in the actual fielded system. Integrating the verification paradigm described above with simulation is a major engineering challenge of the future. Proof strategies should generate test suites; test data should suggest proof approaches.

To be truly meaningful verification must be from top to bottom: conceptual design, specification levels, higher order language, machine code, microcode, gate arrays.

These lectures have a more modest aim. They describe some of the basic theoretical results in program verification; that is, the use of formal systems to prove that software in higher order languages meets its formal specification. We are primarily interested in practical results, results which could be used in next-generation, semantically based, software engineering environments which will support the development of more reliable, dependable, safe systems. We consider a system to be "dependable" (or "predictable") when we know what it will and won't do in any given environment. We view the current proliferation of "unpredictable" software systems as a temporary mark of immaturity. Steam boilers used to frequently explode; today's internal combustion engines explode in a more controlled manner. Rather than constrain capabilities, safety adds new ones.

2.3.1 The Anti-Formal Methods Position

The proposal to infuse software engineering with a large dose of formal methods is not without its opponents. The main emphasis of their opposition is the stubborn inescapable fact that although the theoretical possibility of program verification was established over twenty years ago it is still not a reality in any conventional sense of the word "reality". This is in stark contrast to so much else in computer science – workstations, networking, distributed computing, spectacular graphics, supercomputing, massive parallelism, connectionist machines, etc., etc., – which rapidly transitioned from initial idea to commercial availability. Such progress is breathtaking. Compared with it, the progress in verification is an obvious embarrassment. Too many panels and symposia have been convened both to bemoan the lack of progress and to ask, still again, when can we expect verified software? It doesn't make much sense to ask whether code verification is desirable or to attempt to determine which applications lend themselves to code verification if, in reality, the whole enterprise is not feasible.

As a matter of fact, there has been substantial progress in the field. Why then has this progress been so difficult to discern? The answer seems to be that an inappropriate time scale has been used to measure it. The point will be elaborateed below but it can be summarized in two slogans:

1. Verification is closer to mathematics than it is to other areas of computer science or to engineering.

2. Mathematics, due to its nature, develops at a much slower pace than science or engineering.

In the U. S. there is a weekly magazine called "Science" but it is doubtful whether a weekly "Mathematics" would have much to report. Fundamentally, this is due to the fact that mathematics has no empirical content. It is all thought with no perception. Thought is slow and hesitant compared with perception which is instantaneous and immediate. Perception is a primary method of knowledge in science and technology. One tests or experiments and "sees" what the results are. In mathematics one can only do "thought experiments" where "seeing the results" is replaced by "thinking out the consequences" and the latter takes much more time.

University mathematics teachers quickly discover that among good students a basic distinction can be drawn between those interested in using mathematics for other purposes and those interested in creating and contemplating mathematics for itself. Users, such as engineers, treat mathematics as a tool. They are primarily interested in the mathematical results, the theorems, and tend not to be interested in the proofs of those theorems. It takes much less time to state such results and to teach students how to solve problems using these results than it does to go through a proof. Engineering students are impatient when time is spent expositing proofs because generally proofs don't give them any more tools; they only explain why the tools work and engineers aren't that much interested in why tools work – they want to build. For them, mathematics is problem solving. A very healthy, pragmatic, American point of view. Also a point of view that non-technical people can easily relate to.

Mathematicians, on the other hand, are more interested in proofs than they are in theorems. Why else would they be interested in tens, in some cases hundreds, of proofs of the same result? Mathematicians focus on discovering and contemplating inherent structure, that is, structure which occurs naturally in the mathematical universe. They shy away from anything they feel is "artificial"; indeed in mathematics to say a proof is "artificial" is a severe criticism. It doesn't mean the proof is incorrect. What it does mean is something like "this proof is ugly; it's too ad hoc; it doesn't seem to flow naturally; it's bound to be sterile, that is, it won't lead to any other proofs". For mathematicians the opposite of "artificial" is "elegant." Elegance in mathematics isn't a luxury, a useless add on. On the contrary, elegance is the sign of potency.

Now formulating conjectures and finding and contemplating elegant proofs is a much more time consuming activity than applying theorems to solve problems. The best mathematicians are not necessarily those with the longest publications list: the logician Kurt Gödel wrote about a dozen papers over a thirty year period. Not very prolific. On the other hand, the work in question defined the field. Mathematics moves at a much slower pace than either empirical science or technology. It is not unusual for mathematical conjectures to remain open for decades.

The point is: program verification, and the related formal methods areas such as the semantics of programming and specification languages and automatic theorem proving are closer to mathematics than they are to engineering. Developing graphics workstations, submicron technology, or photonics computers are very different activities than developing verification tools. Engineering tools are adequate exactly when they appear to be: a proposed hammer which hammers is a hammer; graphics software which apparently displays the movements of the mouse actually displays such movements; network software which gives the appearance of transparency is transparent. Indeed, in all these cases it is exactly the appearance which one desires. To get these practical results there is no need to develop a theory before proceeding. In fact, no theory need ever be developed. Consider a boot-strap compiler which is used to compile itself. There is no mathematical theory of what such an object is yet they are built everyday. Verification tools are meant to be used to prove the soundness and correctness of other software. For this exercise to make sense the tools themselves need to be sound and correct in some meaningful manner. At the very least, this requires that the mathematical basis behind the tools, the soundness and completeness of the underlying logics and definitions, be worked out and checked by mathematically competent people. And this takes time, time as measured on the "growth of mathematics" scale not the "growth of technology" scale. Furthermore, the output is much less tangible than ordinary technology efforts. Indeed, a non-sound automatic theorem prover would probably generate more theorems per cpu hour than a sound one. But should you believe them?

One might ask, who needs all this? Everything seems to be working fine without verification. Who needs a proof in addition to the software? The system is its own proof. A purported hammer which hammers is a hammer. Until it doesn't. π was 22/7 until it wasn't, that is, until technology developed sufficient precision that it mattered what the higher decimal places in π were. Verification, or at least something very similar to it, is becoming a necessity. There is an increasingly urgent need for practical validation technology. Note the word "validation". It does not necessarily mean formal verification, particularly when there are near term goals involved. Although there are other definitions of "validation" it is used here to that end of the verification spectrum which is ready for use. It also includes simplifications of verification which are designed to catch well known, commonly occurring errors in a given application area. Information flow analysis is such an example of validation. Flow analysis tools take code as input and yield a table showing for each variable what variables it depends upon. The construction of such tables is a "light weight" form of verification. It does the path analyses described later without proving anything about the program. Nevertheless, if programmers are required to annotate their code with their own expectations of variable dependencies then such a tool can catch discrepancies and bring them to the programmer's attention. In practice this turns out to be a very useful debugging method.

Such validation is a reasonable near-term goal. Furthermore, it is reasonable to believe that the validation area will develop its concepts, methodologies and tools as spin offs from research efforts which take full, formal verification as their goal. As an illustration consider the history of π. Man has used many fractions as an approximation to the ideal π in his engineering accomplishments throughout the ages. The infinite series representation of π has been known for 250 years. One can't actually compute with an infinite series since it is, after all, infinite. But it does provide a uniform way of generating approximating fractions to any degree of precision. Similarly, a sound verification research program should bite the bullet and accept total verification and proof (i.e., top to bottom, from gates through micro-code, assemblers, compilers, operating systems, etc.) as its goal. But it also has an obligation to provide means for generating partial verifications (which we are calling validations) corresponding to various degrees of precision. In fact, this is not only what should be happening this is also what is happening. It just isn't that obvious because one doesn't see the big picture because of a development according to a time scale technology followers are not used to.

In addition to validation tools, formal methods can make an immediate contribution to software engineering in the area of formal specification languages and their use in the development of reusable theories . Historically formal specification languages were primarily developed to provide a language within which one could express the program properties which one was interested in proving. They are now seen to provide a layer of precise documentation useful for recording programmer or designer intentions even in the absence of evidence that these intentions have been reached. They act as a communications media useful both to the present builders and to future maintainers. Most people who have used such formal specifications as part of the software development cycle agree that their role in the clarification of design and the articulation of system/subsystem relations and interfaces is a significant improvement over current natural language practices. In particular, these languages have been used to axiomatize apllication domains like payrolls, bank accounts, book libraries, seat reservations, etc. so that the assumptions under which the programmer is working and the goal he must reach is made explicit. These assumptions and goals might be wrong but it is better to find that out by reviewing formal theories than by having systems fail.

One criticism against formal verification which might be true is that it doesn't seem to scale up. While proving simple facts about short programs is straight forward, proving facts about operating systems is another matter. One rebuttal to this argument is to be equally offensive and ask whether anything scales up? Does anyone understand or have confidence in large systems? Should we trust them? Should we even build them? Such a purist approach ignores the reality of a technological imperative which seems to drive us on in spite of reservations. Rightly or wrongly large, ill understood, systems will be built and will be trusted to some extent in that they will insinuate themselves into our lives. The people responsible for such developments will use the bast available technology to limit the risks but in the forseeable future it won't be verification. Nevertheless, the scientific and technological basis for extending verification to cover large systems is also developing. Mathematical theorems have recently been proved which make essential use of computing machinery in that the proofs themselves are too large and complex to be done by hand. Although philosophers debate whether such "proofs" are really proofs (particularly when the software that carried out the proof was not verified) we are witnessing a remarkable extension of man's theorem proving capabilities. It is natural to expect that the theoretical developments along these lines will soon be available in program verification environments.

2.4 Historical Interlude

An interesting case can be made that deductive mathematics began as a methodology for verifying software. The Greeks inherited a large body of mathematical technique from their Egyptian and Mesopotamian predecessors. Historians are only beginning to understand the extent of such Greek indebtedness. These techniques were essentially arithmetical, algebraic, and geometric algorithms for solving many naturally arising problems. They prescribed, in an imperative manner, a series of steps which started from given data and arrived at a desired outcome. In this way numerical square roots were taken (the Mesopotamian square root algorithm is essentially the same as that which until recently was taught in secondary schools), angles were bisected, perpendiculars were dropped, quadratic equations were solved, and circles were squared.

The techniques came with no accompanying explanation (documentation) as to why they were correct. Indeed, the very notion of "correct" had not yet been defined. The Greeks appear to be the first to distinguish between exact solutions as in angle bisection and approximate solutions as in numerical square root calculation and circle squaring. They were primarily interested in the former. Indeed, reflection on the difference between the two resulted in the formulation of the distinction between "truth" and "opinion"

Figure 1: The Line Segment $\overline{AB}$

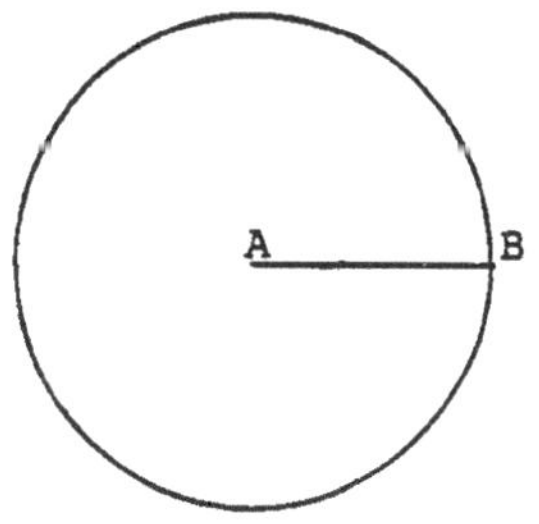

Figure 2: The First Circle

which is perhaps the greatest of Greek achievements.

One could read Euclid's "Elements" as a work on program verification. The underlying primitive operations of the programming language are laid down (drawing a line between two points, extending a line segment, drawing a circle given the center and the radius) and (informal) programs are written in terms of these primitives and proved correct. The proofs use intermediate theorems which are introduced primarily to support the program correctness proofs. What is missing is a consideration of iteration and looping although some of the Eudoxus' theory of irrationals comes very close.

Let us illustrate this interesting theme. One of the oldest complex mathematical operations known to man is the construction of the perpendicular bisector of a line segment. This allows one to both bisect a segment and to raise a perpendicular.

Consider a line segment $\overline{AB}$ as shown in Figure 1. That is the input data. First construct a circle of radius $\overline{AB}$ centered at A as shown in Figure 2. Then construct a circle of radius $\overline{BA}$ centered at B as shown in Figure 3. Let C and D be the points where the two circles intersect. The line segment $\overline{CD}$ which meets $\overline{AB}$ at E is the perpendicular bisector of $\overline{AB}$ as shown in Figure 4.

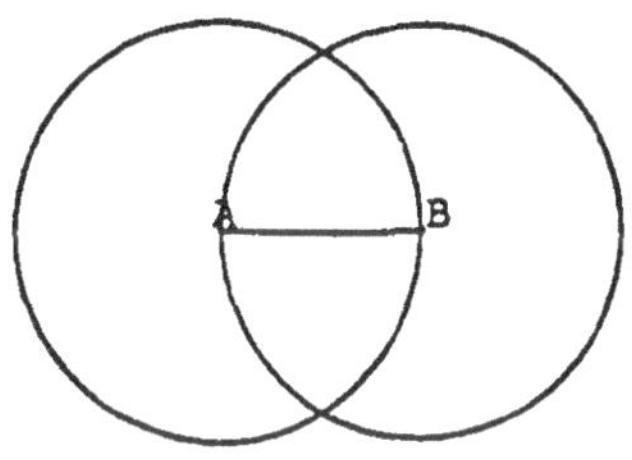

Figure 3: Both Circles

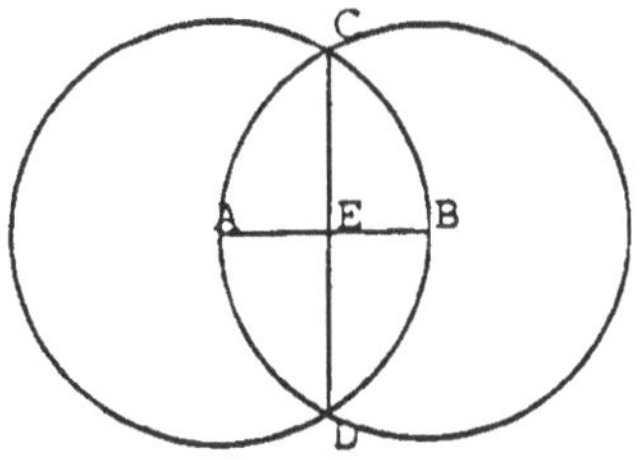

Figure 4: The Perpendicular Bisector

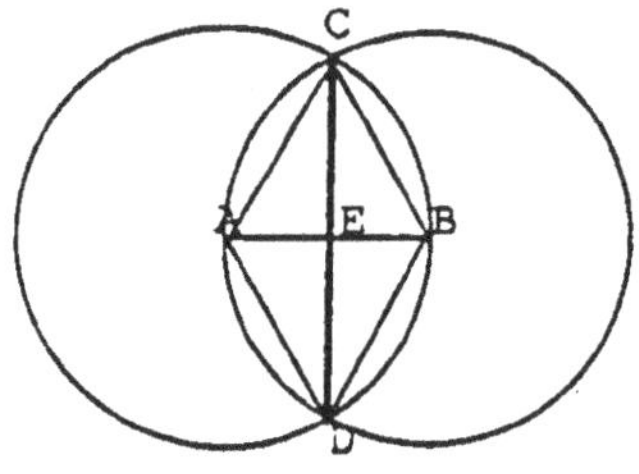

Figure 5: The Underlying Congruent Triangles

Why? If this question had been asked at all prior to the Greeks an acceptable answer might have been "symmetry"; nothing was done on the left that was not done on the right. Hence $\overline{AE}$ is of the same length as $\overline{BE}$ and the $\angle AEC$ has the same magnitude as $\angle BEC$ which is what is meant by saying that $\overline{CD}$ is the perpendicular bisector of $\overline{AB}$.

The Greeks realized that many appeals to "symmetry" can be reduced to the properties of congruent triangles. This was a major mathematical achievement; it remained the method of treating symmetry in geometry until the rise of group theory in the late nineteenth century. All the appropriate triangles are shown in Figure 5.

In our problem everything would follow if one showed that ΔCEA was congruent to ΔCEB. There are several ways to show congruency; the most primary being SAS (side angle side). $\overline{CE}$ is a shared side and $\overline{CA}$ equals $\overline{CB}$ since both equal $\overline{AB}$ since they are radii of circles with $\overline{AB}$ as a radius. Hence it only remains to show that $\angle ACE$ equals $\angle BCE$.

For that one needs to use congruent triangles again. These angles are the corresponding angles in ΔACD and ΔBCD respectively. These two triangles can be shown to be congruent using SSS (side side side) since we've already seen that $\overline{AC}$ equals $\overline{BC}$ and in a similar manner $\overline{AD}$ equals $\overline{BD}$ and $\overline{CD}$ is shared. Alternatively, if one prefers SAS to SSS (since it's more primitive) one can prove ΔACD congruent to ΔBCD by showing that the $\angle CAD$ and $\angle CBD$ are equal. This is done by showing that $\angle CAB$ and $\angle CBA$ are equal (base angles of an isosceles triangle are equal) and then showing that $\angle DAB$ and $\angle DBA$ for the same reason and then adding.

All of the above can be thought of as a conceptualization of the simple appeal to "symmetry". It entails the making conscious of all the underlying reasons. The intellectual level of the proof is more demanding than the construction. One can easily imagine a Mesopotamian wondering why one has to go through all this when it's perfectly clear that the perpendicular bisector has been constructed. It's not as if one gets a new construction when one's through; it's the same old construction just with a lot better documentation (after all the documentation has lasted 2,000 years!).

Analogously, there are those who complain about program verification; if it doesn't help in the discovery of new algorithms or the writing of new programs what good is it? Well, what are the alternatives to proof? Despite all claims to the contrary, Mathematics is, and will always remain, the Queen of the Sciences and from the mathematical point of view, proofs are the only game in town. At various times disciplines like physics have claimed that proofs were too fussy and alternative techniques (most recently it's computer simulation but in the past it's been appeals to "physical intuition") were more appropriate in their field. But the supreme, ruling, value of proof has been repeatedly vindicated. Theoretical physics is more proof based today than ever.

Uncertainty and risk annoy most people who react by declaring things that annoy them to be illegal. The various automobile recall notices previously described came about in the U.S. in response to "The National Trafic and Motor Vehicle Safety Act". It is just a matter of time before similar laws and regulations governing software safety are enacted. What methods will be used to meet various legal requirements?

2.4.1 A Second History Lesson

The above geometrical example does not quite illustrate our point since it is very difficult to know what pre-Greek mathematics really sounded like, that is what discussion accompanied the above construction. A better example is eighteenth century analysis. Although dating from the modern period, that era's mathematics of the infinite appears very strange to modern sensibilities. Consider Euler's derivation of the power series expansion of $\cos x$. From complex analysis we get

$$e^{\imath n\theta} = (e^{\imath\theta})^n$$

which when using polar coordinates is:

$$\cos n\theta + \imath \sin n\theta = (\cos\theta + \imath\sin\theta)^n.$$

Applying the binomial theorem to the right hand side we get:

$$\cos n\theta + \imath \sin n\theta = \sum_{k=0}^{n} \imath^k \binom{n}{k} \cos^{n-k}\theta \sin^k\theta.$$

Now equating the two real components ($\imath^k$ is real exactly when k is even) we get

$$\cos n\theta = \sum_{j=0}^{n/2} (-1)^j \binom{n}{2j} \cos^{n-2j}\theta \sin^{2j}\theta.$$

So far, so good. Now Euler lets θ be infinitely small and n be infinitely large. According to Euler, this makes $x = n\theta$ finite, $\sin\theta = \theta$, and $\cos\theta = 1$. Furthermore, we have

$(n-k)\theta = n\theta - k\theta = n\theta = x$ for finite k since $k\theta$ is infinitely small. Hence,

$$\binom{n}{2j}\theta^{2j} = \frac{n(n-1)\dots(n-2j+1)}{(2j)!}\theta^{2j} = \frac{n^{2j}}{(2j)!}\theta^{2j} = \frac{x^{2j}}{(2j)!}$$

We thus get:

$$\cos x = \sum_{j=0}^{\infty}(-1)^j\frac{1}{(2j)!}x^{2j}$$

which, lo and behold, is the correct power series expansion of $\cos x$. Of course, if one makes other substitutions in a similar style then wrong formulas would result. The point is Euler was a "good" mathematician in the eighteenth century sense. He very rarely got wrong formulas. But God knows why. Explicitly or implicitly, he had his rules and constraints which governed the kinds of substitutions he made. But his texts do not reveal them. If we tried to work in his style, we would make a botch of it. Although Euler's results remain, his methodology was swept away by nineteenth century reforms. It was replaced by a method that is much more conscious and explicit, although also more pedantic. The pedantry is the cost of progress; it represented the transition of Mathematics from a Romantic to a Professional era. It should be pointed out that Euler's proof can be rehabilitated using non-standard analysis; in some places "=" needs to be replaced by "infinitely close".

Now for our point. Unconstrained programming in the wild is a bit like Euler's style. In the hands of a master, it works. But masters are few. The nineteenth century "epsilon-delta" analysis proofs, while sometimes unbearably painful, protect the discipline. It is harder for false results to get into circulation. What is being proposed by logically minded computer scientists is a revolution in programming style on the order of the nineteenth century revolution in analysis. After such intellectual revolutions there are few traces left of the "ancien régime". Needless to say, there are strong reactionary tendencies opposing the introduction of formal methods. These tendencies resort to all sorts of attacks including large doses of anti-intellectualism formulated as pragmatism.

3 Flowcharts

A classic definition of a program is that it consists of control structures and data structures. Even if recent developments have extended the notion of program (e.g., functional programming, object-oriented programs, spreadsheets) the classic definition still holds at some underlying level. Data structures have fairly well known mathematical analogues; indeed abstract algebra has played a significant role in the design of modern programming languages. What is new in computer science, in contrast to mathematics, is the notion of locus of control during a computation. We thus begin our investigations by ignoring the structure of our data and focusing on control. We begin with the most primitive form of notation for control structure — the flowchart.

3.1 Flowchart Schemas

We will assume a signature τ which provides the framework for a language $\mathcal{L}(\tau)$ (or just $\mathcal{L}$ when τ is fixed). τ is any multi-sorted signature which provides a finite set of constant, function, and relational symbols, including equality, each having an associated degree and sort specification. $\mathcal{L}$ is the weak second order language over τ. In particular, it contains additional sorts: **nat** for the natural numbers and for each of τ's sorts a new sort for finite sequences over that sort. In addition, the language contains appropriate constants, functions, and relations for speaking about these new sorts. We needn't be too precise at this point. $\mathcal{L}$ is closed under the propositional connectives and quantifiers over the sorts of τ, **nat**, and the finite sequences. The language is called second order because of the quantifiers over sequences. It is called weak second orer because the sequences are finite.

Since we are focusing on control, in contrast to data, we will assume τ contains only a single sort and ignore typing issues. In addition to the above, a flowchart schema over τ needs program variables. Our formalism can be extended so that each program can declare its variables but for now they will be taken from the list $x_1, x_2, \ldots$; a given program will use the first n of these for some fixed n. We write $\vec{x}$ for the vector of program variables $(x_1, \ldots, x_n)$. Although syntactically all variables look alike there is a distinction between logical variables whose values depend on an interpretation and program variables whose values depend on both interpretation and state. This distinction will become clearer when we discuss semantics. We postpone to the second part of this work a discussion of programs which use the finite sequences in computation (usually called arrays in computer science).

A flowchart schema $\mathcal{F}$ over $\mathcal{L}$ is a finite, labeled directed graph. A directed graph consists of nodes and arrows between nodes. An arrow which leaves a node is called an out arrow with respect to that node; an arrow which enters a node is called an in arrow with respect to that node. A flowchart has five kinds of nodes:

- START: There is a unique START node, it has no in arrows and a unique out arrow.
- STOP: A STOP node has no out arrows.
- ACTION: An ACTION node has a unique out arrow.
- DECISION: A DECISION node has two out arrows; one labeled *true*, the other *false*.
- SKIP: A SKIP node has a unique out arrow.

Each ACTION node is labeled with an assignment statement of the form

$$x_i := e$$

where e is an expression built up from the constants and functions in τ using the variables in $\vec{x}$. Each DECISION node is labeled with a formula $\phi(\vec{x})$ from $\mathcal{L}$ whose free variables are among $\vec{x}$. $\phi(\vec{x})$ is taken from the quantifier-free part of $\mathcal{L}$; that is, it's a propositional or boolean combination of atomic formulas including equalities. Note that allowing such ϕ to occur at DECISION nodes reveals our assumption that in all applications the predicates, functions and constants of τ will be interpreted by computable objects. A similar remark holds with respect to using formal τ terms in assignment statements.

3.2 Operational Semantics

The programs we have been considering are really flowchart schemas; the constant, function, and relational symbols occurring within them are syntactic, having no intrinsic meaning. A flowchart is an interpreted schema and its operational semantics is defined with respect to an interpretation $\mathcal{A}$. $\mathcal{A}$ provides a "universe of discourse", $|\mathcal{A}|$, which is a non-empty set (the set of values of the single sort over which the program is written) and interpretations of the appropriate kind for each constant, function, and relation symbol of τ. Given a flowchart schema $\mathcal{F}$, an interpretation $\mathcal{A}$, and a choice of initial values $\vec{a}$ (also called $\vec{x}_{init}$) from $|\mathcal{A}|$ for the program variables $\vec{x}$, the run $\mathcal{F}(\vec{a})$ of the program begins at the unique START node and steps through $\mathcal{F}$ taking the appropriate actions at each ACTION node and following the appropriate out arrow at each DECISION node. There is no action at a SKIP node; control just passes through. For the purposes of semantics, we can think of SKIP nodes as ACTION nodes with the no-op assignment

$$x_1 := x_1.$$

SKIP nodes are only used for convenience in describing in line expansions. We postpone for now the question of how the initial values of the variables are actually passed into the program variables; instead we concentrate on the run.

A more mathematical definition of the run $\mathcal{F}(\vec{a})$ can be given in terms of states and transition rules. A state is a pair consisting of the current value of the program variables $\vec{x}_{current}$ and the current node of $\mathcal{F}$. The initial state consists of $\vec{a}$ together with the unique START node. Transition rules correspond to the kind of node one is currently at. At an ACTION node $\vec{x}_{current}$ is changed by performing the appropriate assignment $x_i := e$. Here, e is evaluated using $\vec{x}_{current}$ as the values of the program variables. After e has been evaluated, $\vec{x}_{current}$ is changed and the current node is changed to the target of the given ACTION node's unique out arrow. At a DECISION node the truth value of the formula $\phi(\vec{x})$ is calculated using $\vec{x}_{current}$ for the values of the free variables $\vec{x}$ and the current node is changed appropriately by following either the *true* or *false* out arrow. The value of $\vec{x}_{current}$ is not changed. This approach defines the operational semantics in terms of rewrite rules. The initial state is continually rewritten using the transition rules. The rewrite rules described above are deterministic; only one transition rule can apply to a state.

The above discussion also shows what kind of mathematical objects program variable are. Consider the assignment $x_i := e$. In order to evaluate e in state s one has to read the values of the $\vec{x}$ at s. Hence each x_j is a function from states to values (using the representation of states as pairs of current values and program location we see that each x_j is a projection function). Expressions e are also maps from states to values. We can write $x_j s$ and $e[s]$ for the values of these functions at state s, On the other hand the x_i on the left side of the assignment plays a different role. The assignment is a state transition function, T_i which maps states s into states s'. Furthermore the transition T_i is defined by

$$x_j[T_i(s)] = \text{if } (j = i) \text{ then } e_i[s] \text{ else } x_j[s].$$

Returning to the operational semantics we say that the run halts if it ever reaches a STOP node; we will consider the case of non-terminating runs later. The value of $\vec{x}_{current}$

when the run halts is called $\vec{x}_{final}$. Since $\mathcal{F}(\vec{a})$ might not halt $\vec{x}_{final}$ might not be defined for a given $\vec{x}_{init}$. The relationship between the initial and final values of $\vec{x}$ will be denoted by $\Phi_{\mathcal{F}}(\vec{x}_{init}, \vec{x}_{final})$. When considering a fixed flowchart the subscript $\mathcal{F}$ will be dropped and for convenience $\vec{x}_{init}$ will be written as $\vec{y}$ and $\vec{x}_{final}$ will be written as $\vec{z}$ so that for a given flowchart the basic relationship appears as:

$$\mathcal{A} \models \Phi(\vec{y}, \vec{z})$$

which is read as

In interpretation $\mathcal{A}$, $\mathcal{F}$ terminates when started with $\vec{y}$ with resulting value $\vec{z}$.

Our discussion provides a semantic definition of Φ but we would like to know its syntactic form. In particular since $\mathcal{F}$ is schematic we would like Φ to be schematic also; that is, we would like to find a syntactic formula for Φ which is uniform for all interpretations $\mathcal{A}$. Unfortunately, Φ might not be expressible in $\mathcal{L}$. Note that the variables $\vec{y}$ and $\vec{z}$ which appear in Φ are "logical" variables, that is variables that would occur in formulas of $\mathcal{L}$ in contrast to the "program" variables $\vec{x}$ which occur in $\mathcal{F}$ which are, as we have seen, state functions. This distinction between "logical" and "program" variables is frequently a source of confusion. The confusion is understandable since the formulas $\phi(\vec{x})$ at DECISION nodes use program variables as if they were logical variables. In reality, $\phi(\vec{x})$ is not a formula of $\mathcal{L}$ since its truth value depends on the state of the run. One of the main conceptual attractions of functional programming languages (in contrast with the imperative program language we are describing) is that only logical variables occur.

3.2.1 Non-Determinism

The various analyses of Φ which we give will make very little use of the deterministic nature of $\mathcal{F}$. In particular, we can relax the condition that the START and ACTION nodes have unique out arrows and that DECISION nodes have unique *true* and *false* out arrows. We can also relax the condition that DECISION nodes have unique formulas attached to them. We allow each DECISION node to have an attached finite sequence $(\phi_1, \ldots, \phi_k)$ of $\mathcal{L}$ formulas. In addition, at each such node there will be at least one $true_i$ out arrow for each ϕ_i plus at least one *false* out arrow.

The transition rules are then non-deterministic; that is, in a given state there is more than one state one can transition to: one can leave START and ACTION nodes along one of several arrows and at a DECISION node if some ϕ_i is true then one can leave along one of its $true_i$ out arrows otherwise one leaves along one of the *false* out arrows. If several ϕ_i are true at a DECISION node the computation chooses one. In summary: the unique START node has no in arrows and at least one out arrow; STOP nodes have no out arrows; ACTION nodes have at least one out arrow; DECISION nodes have at least one $true_i$ for each attached ϕ_i and at least one *false* out arrow. Since for semantics sake we are treating SKIP nodes as ACTION nodes (with a no-op action) we see that SKIP nodes must have at least one out arrow. As a further element of non-determinism we admit the non-determinisitc assignment

$$x_i := ?$$

which is executed in a given run by assigning an arbitrary element of the value domain to the variable.

The correctness relation

$$\mathcal{A} \models \Phi(\vec{y}, \vec{z})$$

means that with input data $\vec{y}$ from $|\mathcal{A}|$ there is some computation of $\mathcal{F}$ which terminates with final value $\vec{z}$; for a fixed $\vec{y}$ there may be several $\vec{z}$ (even infinitely many) which satisfy Φ. After defining Φ the rest of the discussion for non-deterministic flowcharts just repeats the discussion for deterministic flowcharts and we assume that has been done.

Non-determinism has many uses. In particular, it is used in one of the standard operational semantics of parallelism. Here several programs with common program variables run in parallel. At each step only one can proceed but which is non-determined. This model of parallelism is called "interleaving".

3.3 Straightline Flowcharts

The logical form of Φ is easy to determine if there are no loops in the directed graph $\mathcal{F}$. A loop is a path which begins and ends on the same node; a path is a sequence of nodes linked by arrows. In a loopless flowchart there are only finitely many paths between any two nodes. Hence there are only finitely many paths that begin at the START node and end at some STOP node.

Let us assume that there are no non-deterministic assignments in the flowchart. Then for each path P from START to a STOP it is possible to symbolically execute the program as one moves down P; that is, starting with the formal expression $\vec{y}$ one can perform the appropriate updates $x_i := e$ at each ACTION node on P. The result will be some vector of terms $\vec{t}_P$ formed from the logical variables $\vec{y}$ and the constants and function symbols of τ. Furthermore, we can construct an $\mathcal{L}$ formula ϕ_P with $\vec{y}$ as its free variables which corresponds to the path P. This is done by taking the conjunction of the appropriately instantiated decision formulas ϕ's or their negations at each of P's DECISION nodes. For example, at a non-deterministic DECISION node the formula correspond to a *false* out arrow is

$$\neg\phi_1 \wedge \ldots \wedge \neg\phi_k$$

In the deterministic case ϕ_P is true if and only if the given path is followed during the computation starting with $\vec{y}$. In the non-deterministic case ϕ_P is true if and only if there is some computation with which starts with $\vec{y}$ and follows P. The path P is fully reflected by the formula

$$\phi_P(\vec{y}) \wedge \vec{z} = \vec{t}_P$$

and Φ is the finite disjunction of these, one for each path P from START to a STOP.

If we now allow non-deterministic assignments then we alter our description of how the the symbolic execution is updated by introducing a new logical variable for the value of the program variable being assigned to at such assignments. The resulting term $\vec{t}_P$ now not only contains the logical variables $\vec{y}$ but also new variables $\vec{w}$ and similarly the path predicate ϕ_P depends on both $\vec{y}$ and $\vec{w}$. We then apply existential quantifiers to bind all such new logical variables. The resulting formula is the correct semantics; it says that some value was chosen at each non-deterministic assignment statement encountered along the path. The path correctness predicate then becomes

$$\exists \vec{w}[\phi_P(\vec{y},\vec{w}) \wedge \vec{z} = \vec{t}_P(\vec{y},\vec{w})].$$

As we shall see in the second part of this work, non-deterministic assignment is a useful mechanism for modeling the declaration of local variables which before explicit assignment can be conceived of as having a value assigned by a non-deterministic assignment.

3.4 Unwinding

The above analysis breaks down in the presence of loops. Loops can be removed if the flowchart is unwound. The result is, unfortunately, an infinite tree.

Let $\mathcal{F}$ be a fixed flowchart. We now form its tree $\mathcal{T}_{\mathcal{F}}$ or $\mathcal{T}$ for short. The nodes of $\mathcal{T}$ are the finite paths of $\mathcal{F}$ which begin with START. An arrow joins two nodes of n_1 and n_2 of $\mathcal{T}$ if as paths in $\mathcal{F}$ n_2 is consists of the $\mathcal{F}$ path n_1 followed by one more $\mathcal{F}$ node. $\mathcal{T}$ nodes and arrows are labeled using the labels attached to the corresponding $\mathcal{F}$ nodes and arrows. Clearly, $\mathcal{T}$ is infinite if $\mathcal{F}$ has at least one loop.

To each terminating path P in $\mathcal{T}$ one can assign a formula as before which states that starting with $\vec{y}$ and making some assignments at the non-deterministic assignment nodes the path P was traversed and $\vec{z}$ is the resulting value. $\Phi(\vec{y},\vec{z})$ is now the countable disjunction of these formulas. We thus have shown:

Theorem 1 *Φ is in $\mathcal{L}_{\omega_1,\omega}$ over $\mathcal{L}$.*

As usual in logic, $\mathcal{L}_{\omega_1,\omega}(\tau)$ is the infinitary language over τ which in addition to the usual propositional and quantifier closure rules allows countable disjunctions and conjunctions. This classification of Φ is very crude since, as the proof shows, Φ is actually a countable disjunction of finite conjunctions of atomic formulas and their negations (and existential quantifications if non-deterministic assignments are allowed). The aim of the rest of this report is to show that Φ is actually in the weak second order language over τ.

4 Inductive Definitions

As we mentioned, the result that Φ is $\mathcal{L}_{\omega_1,\omega}$ is unsatisfactory since the infinitary operations are hardly used in forming Φ. Corresponding to each loop there is a countable disjunction which essentially corresponds to going around the loop once, twice, etc. Of course in flowchart programs, loops can be arbitrarily intertwined. The correctness relation Φ takes the form it does since it is a direct encoding of the operational semantics of the flowchart.

In the early days of programming language theory John McCarthy pointed out that the semantics of imperative languages could be more cleanly described using recursion theory. One can define a set of mutual recursion equations which abstract away from the actual stages and states of the computation. This remark led to the whole area of functional programming (first in Lisp, then ML, FP, Miranda, Haskell, etc.) in which one programs at a higher level, namely in a metalanguage adequate for formalizing the semantics of imperative languages.

Alternatively, once can formulate the semantics of imperative languages in the theory of inductive definability. Analogous to recursive semantics this also leads to a new style of programming, logic programming. It also yields a very convenient formulation of the correctness relation Φ as well as a method for proving the correctness of imperative programs. Although our results are classical (Floyd's Inductive Assertion Method) we believe our approach is novel but this may be due to our ignorance of the computer science literature. Anyway, as a logician we find it quite satisfying.

Using inductive definability for semantics instead of functional recursion equations is coherent with the extensive use of such definitions throughout computer science. The syntax of programming languages, for example, is given in terms of inductive definitions of expressions, statements, and program units. Computer scientists sometimes refer to such definitions as recursive but they really mean inductive.

We begin by reviewing the well known results about inductive definitions and then apply these facts to Φ.

4.1 Inductive Definitions as Least Fixed Points

Explicit definitions of subsets A of a base set X have the form

$$x \in A \Leftrightarrow \chi(x)$$

where $\chi(x)$ is a logical formula not containing A and the variable x ranges over X. Implicit definitions, on the other hand have the more general form

$$x \in A \Leftrightarrow \chi(x, A). \tag{1}$$

Such definitions are self-referential; the object A being defined is referred to within its own definition. Self-referential definitions appear paradoxical. Indeed, since A is allowed to appear in χ care must be taken since not every purported definition of this form is meaningful (for example, let $\chi(x, A)$ in formula (1) be "$\neg(x \in A)$".) Definability theory is that area of logic which investigates the justification of various forms of definitions.

Inductive definitions are special cases of implicit definitions. They either take the form: A is the least subset of X such that

$$x \in A \Leftrightarrow \chi(x, A)$$

or the form: A is the least subset of X such that

$$\chi(x, A) \Rightarrow x \in A \tag{2}$$

which appears more like a closure condition than a definition. As we shall see these two forms define the same set when χ-is monotone. Monotoness is the simplest restriction on χ which guarantees the existence of the inductively object.

Definition 1 *A logical formula* $\chi(x, A)$ *is* **monotone** *in* A *if for all* $x \in X$

$$[B \subseteq C \wedge \chi(x, B)] \Rightarrow \chi(x, C).$$

Note that $\neg(x \in A)$ is not monotone in A.

$\chi(x, A)$ is formula (2) is usually a disjunction $\bigvee_{i=1}^{n} \psi_i(x, A)$ so that formula (2) becomes: A is the least subset of X such that:

$$\psi_1(x, A) \Rightarrow x \in A$$

$$\vdots$$

$$\psi_n(x, A) \Rightarrow x \in A.$$

The case of a single inductive definition can be generalized to mutual inductive definitions of the form: A, B are the least sets such that

$$x \in A \Leftrightarrow \chi_1(x, A, B)$$

$$x \in B \Leftrightarrow \chi_2(x, A, B)$$

where the $\chi_i(x, C, D)$ are monotone in both C and D. This case reduces to the simple case when the problem is treated abstractly as we do.

Another generalization is *relative* inductive definitions. Consider the definition of the ancestral, $*R$, of a binary relation R. It's the least transitive relation containing R and can be defined by: $*R$ is the least relation S such that

$$R(x, y) \Rightarrow S(x, y)$$

$$[S(x, z) \wedge S(z, y)] \Rightarrow S(x, y).$$

More generally, given a logical formula $\chi(x, A, B)$ which is monotone in A an operator $T(B)$ can be defined by: $T(B)$ is the least set such that

$$x \in T(B) \Leftrightarrow \chi(x, T(B), B).$$

From now on we restrict ourselves to the unrelativized case but the reader can see that the relativized case follows by carrying along the extra parameter.

To simplify our discussion of we introduce the function

$$\Lambda_\chi(B) = \{x | \chi(x, B)\}.$$

Then the A defined by formula (1) is the least fixed point of Λ_χ and the set defined by formula (2) is the least B with

$$\Lambda_\chi(B) \subseteq B.$$

The function Λ_χ maps the power set of X, $\mathcal{P}(X)$, into itself. The obvious logical questions are: when does Λ_χ have fixed points; when are they unique; and if they are not unique when are there distinguished fixed points such as a least?

4.2 Inductive Definitions Based on Iterations

In order to discuss such inductive definitions it turns out that it is more convenient to consider maps Λ defined not on all of the power set, $\mathcal{P}(X)$ but on inductive subcollections.

Definition 2 *A subset $\mathcal{C}$ of $\mathcal{P}(X)$ is called* **inductive** *if whenever $\mathcal{B}$ is a subset of $\mathcal{C}$ which is linearly ordered by $\subseteq$ then $\bigcup \mathcal{B} \in \mathcal{C}$*

$\mathcal{B}$ linearly ordered by $\subseteq$ means:

$$E, D \in \mathcal{B} \Rightarrow [E \subseteq D \vee D \subseteq E].$$

Note that inductive collections are non-empty since they always contain $\emptyset$ which is $\bigcup \emptyset$.

Definition 3 *Suppose Λ is a map on an inductive $\mathcal{C}$. Λ is* **monotone** *if*

$$B \subseteq C \Rightarrow \Lambda(B) \subseteq \Lambda(C).$$

Λ is **expansive** *if*

$$B \subseteq \Lambda(B).$$

The localization to inductive collections allows us to reduce simultaneous definitions to simply ones. For example, given formulas $\chi_i(x, C, D)$ monotone in C and D then

$$\Lambda_\chi(C, D) = \{(x, y) | \chi_1(x, C, D) \wedge \chi_2(y, C, D)\}$$

is not defined on all of $\mathcal{P}(X \times X)$ but only on the inductive

$$\mathcal{C} = \{C \times D | C \subseteq X \wedge D \subseteq X\}$$

on which it is monotone.

Although we are primarily interested in monotone operators it is easier to make the construction using expansive operators.

Theorem 2 *Suppose Λ is a monotone map on the inductive $\mathcal{B}$. Let*

$$\mathcal{C} = \{B \in \mathcal{B} | B \subseteq \Lambda(B)\}.$$

Then $\mathcal{C}$ is a non-empty, inductive subcollection of $\mathcal{B}$, closed under Λ on which Λ is monotone and expansive. Furthermore, all fixed points of Λ are in $\mathcal{C}$.

Proof of Theorem: Clearly the fixed points of Λ are in $\mathcal{C}$. Furthermore, $\mathcal{C}$ is non-empty since $\emptyset$ is in it. To show $\mathcal{C}$ is inductive suppose $\mathcal{A}$ is a subcollection of it linearly ordered by $\subseteq$ with $A = \bigcup \mathcal{A}$. We have to show $A \subseteq \Lambda(A)$. But if $B \in \mathcal{A}$ then $\Lambda(B) \subseteq \Lambda(A)$ since Λ is monotone. But Λ is expansive at B, so $B \subseteq \Lambda(A)$. Hence, $\Lambda(A)$ is an upper bound of $\mathcal{A}$ so

$$\bigcup \mathcal{A} = A \subseteq \Lambda(A).$$

Lastly we have to show that $\mathcal{C}$ is closed under Λ. But since Λ is monotone $B \subseteq \Lambda(B)$ implies

$$\Lambda(B) \subseteq \Lambda(\Lambda(B))$$

so that $\Lambda(B)$ is in $\mathcal{C}$ when B. By definition, Λ is expansive on $\mathcal{C}$ and it's also monotone. QED

We now define the basic iterative construction of fixed points. For generality we make use of the ordinals but in all our applications to programs we will need only finite ordinals (i.e., the natural numbers) so the reader can ignore all mention of transfinite notions like limit ordinals.

Suppose Λ is an expansive map on an inductive $\mathcal{B}$ and suppose $A \in \mathcal{B}$. We define a sequence $\Lambda_\alpha(A)$ by transfinite recursion over the ordinals:

$$\Lambda_0(A) = A$$

$$\Lambda_{\alpha+1}(A) = \Lambda(\Lambda_\alpha(A))$$

and

$$\Lambda_\lambda(A) = \bigcup\{\Lambda_\beta(A) | \beta < \lambda\}$$

when λ is a limit ordinal. Since $\mathcal{C}$ is not closed under arbitrary unions one has to show that the collection over which the union is being taken is linearly ordered by $\subseteq$. One does this by proving by ordinal induction that

$$\alpha < \beta \Rightarrow \Lambda_\alpha(A) \subseteq \Lambda_\beta(A).$$

Since $\mathcal{C}$ is a subset of $\mathcal{P}(X)$ we can not have proper inclusion

$$\Lambda_\alpha(A) \subset \Lambda(\Lambda_\alpha(A))$$

for all α. Hence there must be a least α with

$$\Lambda_\alpha(A) = \Lambda(\Lambda_\alpha(A)).$$

We denote such $\Lambda_\alpha(A)$ by $\Lambda^*(A)$. It is a fixed point of Λ which contains A.

Suppose in addition to expansive that Λ is also monotone. If $A \subseteq B$ and B has the property $\Lambda(B) \subseteq B$. Then

$$\Lambda^*(A) \subseteq B.$$

This is shown by proving

$$\Lambda_\alpha(A) \subseteq B$$

by induction on α. Suppose the statement holds for α. Then since Λ is monotone we have

$$\Lambda(\Lambda_\alpha(A)) \subseteq \Lambda(B)$$

which when combined with $\Lambda(B) \subseteq B$ yields

$$\Lambda_{\alpha+1}(A) \subseteq B.$$

This shows that if Λ is both monotone and expansive then $\Lambda^*(A)$ is the least fixed point of Λ containing A. It is also the smallest superset B of A with the property that $\Lambda(B) \subseteq B$.

Combining this construction with the previous Theorem we get:

Theorem 3 *Suppose Λ is a monotone map on the inductive $\mathcal{B}$ and A has the property that $A \subseteq \Lambda(A)$. Then $\Lambda^*(A)$ defined above is both*

1. *The least fixed point of Λ containing A; and*
2. *The least B containing A with $\Lambda(B) \subseteq B$.*

We thus have as a Corollary:

Corollary 4 *Suppose Λ is a monotone operator on the inductive $\mathcal{B}$. Then Λ has a least fixed point. This least fixed point is also the least set B with $\Lambda(B) \subseteq B$.*

The least fixed point is $\Lambda^*(\emptyset)$. Actually our constructions yield more. Namely, suppose Λ is a monotone operator on the power set $\mathcal{P}(X)$ and $\mathcal{C}$ is any collection of fixed points of Λ. Let $A = \bigcup \mathcal{C}$. Then Λ is expansive at A (this requires a lemma to the effect that monotone operators are expansive on unions of sets on which they are expansive) so that $\Lambda^*(A)$ is a fixed point of Λ. It is the smallest fixed point of Λ containing all the sets in $\mathcal{C}$. In particular, note that if $\mathcal{C}$ is all the fixed points of Λ then the set constructed is the largest fixed point.

4.3 Inductive Assertion Method

Suppose A is inductively defined using the monotone $\chi(x, A)$ and we would like to prove some property $P(x)$ for all $x \in A$. The following theorem provides the justification for a method which is the key to showing that programs have specific properties.

Theorem 5 *Suppose*

$$\chi(x, \{y | P(y)\}) \Rightarrow P(x)$$

where χ is a monotone formula inductively defining the set A. Then

$$x \in A \Rightarrow P(x).$$

The result follows from the fact that formula (5) implies that $B = \Lambda_\chi(\{y|P(y)\})$ satisfies the condition

$$\Lambda_\chi(B) \subseteq B$$

and A is the least set which satisfies this condition. We call the method of proof described by this theorem the "Inductive Assertion Method".

We state it more generally. Suppose the relations $R_i(x_1, \ldots, x_{n_i})$ are simultaneously inductively defined by being the least relations satisfying

$$R_i(x_1, \ldots, x_{n_i}) \Leftrightarrow \chi_i(x_1, \ldots, x_{n_i}, R_1, \ldots, R_m)$$

where $i = 1, \ldots, m$ and the χ are monotone. Suppose we wish to prove for a given relation Q_m that

$$R_m(x_1, \ldots, x_{n_m}) \Rightarrow Q_m(x_1, \ldots, x_{n_m}).$$

This can be accomplished if we can find relations Q_i, $i = 1, \ldots, m$ with

$$\chi_i(x_1, \ldots, x_{n_i}, Q_1, \ldots, Q_m) \Rightarrow Q_i(x_1, \ldots, x_{n_i})$$

are all provable for $i = 1, \ldots, m$. These statements are called the "verification conditions". The utility of this result is that the given Q_m and the other Q_i are usually syntactically simpler than the inductively defined R_i. In particular, as we shall see, the latter are weak second order formula (if the χ_i are and are also finitary) while the Q can be usually taken to be first order. If in addition the χ are themselves first order then the verification conditions are all first order and while there are complete systems for first order logics, second order systems must be incomplete.

Note that the "verification conditions" result for provability just relies on the fact that for monotone Λ the least fixed point is also the least B with

$$\Lambda(B) \subseteq B.$$

4.4 Finite Iterations

When do we not need to go beyond the natural number in the basic iterative construction? That is, when is $\Lambda_\omega(A)$ a fixed point? It would suffice to know that Λ preserved the limit of increasing ω sequences, that is if there were a sequence of sets B_i with

$$i < j \Rightarrow B_i \subseteq B_j$$

then

$$\Lambda(\bigcup\{B_i | i < \omega\}) = \bigcup\{\Lambda(B_i) | i < \omega\})$$

for then for any A we would have

$$\Lambda(\Lambda_\omega(A)) = \Lambda(\bigcup\{\Lambda_i(A) | i < \omega\})$$

$$\Lambda(\bigcup\{\Lambda_i(A) | i < \omega\}) = \bigcup\{\Lambda(\Lambda_i(A) | i < \omega\}$$

$$\bigcup\{\Lambda(\Lambda_i(A) | i < \omega\} = \bigcup\{\Lambda(\Lambda_{i+1}(A) | i < \omega\}$$

$$\bigcup\{\Lambda(\Lambda_{i+1}(A) | i < \omega\} = \Lambda_\omega(A).$$

A condition which guarantees that Λ preserves the union of an increasing ω sequence of sets is continuity.

Definition 4 *A map Λ on a power set $\mathcal{P}(X)$ is* **continuous** *if*

$$x \in \Lambda(A) \Leftrightarrow \exists B[B \subseteq A \wedge B \textit{ finite} \wedge x \in \Lambda(B)].$$

The condition says that continuous set maps are those for which membership in the value set depends only on a finite subset of the argument set. On the basis of this Theorem continuous set maps are also called **finitary**.

4.5 Applications to Inductive Definitions

There is a simple syntactic check which guarantees that a formula χ is finitary. Suppose $\chi(x, A)$ is a formula of weak second order logic in which A occurs positively and essentially only within the scopes of existential first-order and weak second-order quantifiers then χ is finitary in A. A subformula occurs positively (negatively) within a larger formula if it occurs within an even (odd) number of negations assuming all propositional connectives are written out in terms of conjunction, disjunction and negation. To say A is essentially only within the scopes of existential quantifiers means that if χ has a subformula α of the form $\forall z\beta(x, A)$ actually containing A then A occurs negatively in β and α occurs negatively in χ. We call such first-order formulas "finitary" in A.

Suppose we are given a finitary χ. How can we use it to determine the form of the resulting inductively defined set? We just apply the discussion above. Let $\chi_0(x)$ be the formula *false* and if $\chi_n(x)$ has been defined then let $\chi_{n+1}(x)$ come from $\chi(x, A)$ by replacing all subformulas in it of the form $t \in A$ by $\chi_n(t)$. The set inductively defined by χ is given by the infinite disjunction:

$$\chi_0(x) \vee \chi_1(x) \vee \chi_2(x) \vee \chi_3(x) \vee \ldots$$

Thus if the first-order χ is finitary then the set it inductively defines is in a $\mathcal{L}_{\omega_1,\omega}$ form. We met this formula earlier when describing the correctness predicate Φ.

4.6 Weak Second Order Logic

The infinitary and second order forms for inductive definitions are much too crude, especially when the defining conditions are finitary. In this section we introduce the language of weak second order logic which is adequate to explicitly express such inductive definitions. If s is a finite sequence then $lh(s)$ is the length of s, if $i < lh(s)$ then $s(i)$ is the i-th component of s, $\overline{s}$ is the set $\{s(i) | i < lh(s)\}$, and $\overline{s}(i)$ is the set $\{s(j) | j < i\}$ if $i < lh(s)$.

Our choice of language is based on the following theorem in which i ranges over natural numbers.

Theorem 6 *Suppose $\chi(x, A)$ is finitary. Then the set it inductively defines has the explicit form:*

$$\exists s \forall i < lh(s)[\chi(s(i), \overline{s}(i)) \wedge \chi(x, \overline{s})] \tag{3}$$

Proof of Theorem: Let B be set inductively defined by χ. We first show that if x satisfies formula (3) then its in B. We prove this by induction on the length of the sequence s which witnesses that x satisfies (3). Suppose s satisfies the body of (3) and that y is in B whenever there is a sequence shorter than the length of s which witnesses for y. Under this assumption we see that

$$\overline{s} \subseteq B$$

but then

$$\chi(x, B)$$

by monotonicity of χ and the fact that

$$\chi(x, \overline{s}).$$

Since B is inductively defined by χ this shows

$$x \in B.$$

For the converse assume $x \in B$ and we will show that x satisfies (3). By the finitariness of χ we know that x is in some B_n where

$$B_0 = B_{j+1} = \{z | \chi(z, B_j)\}.$$

We show x satisfies (3) by induction on n. If n is 0 then x can not be in B_0. Suppose that $x \in B_{j+1}$ so that

$$\chi(x, B_j).$$

By the finitariness of χ there is a finite subset C of B_j with

$$\chi(x, C).$$

By the induction assumption for each y in C there is a finite sequence s_y with

$$\exists s_y \forall i < lh(s_y)[\chi(s_y(i), \overline{s_y}(i)) \wedge \chi(y, \overline{s_y})].$$

Let s be the concatenation of all s_y followed by a sequence which lists C. It is straightforward to see that this s witnesses that x satisfies (3). QED

Note that

$$\chi(z, \overline{s}(w))$$

means that whenever $t \in A$ occurs in $\chi(z, A)$ it is replaced by

$$\exists j < w(s(j) = t).$$

The above discussion was informal. We now assume that ordinary first order logic over τ (which we have restricted to being of a single sort which we now call **k**)is extended with a two new sorts. The first is the sort of natural numbers (called **nat**) and the second is the sort of finite sequences of elements of sort **k**. The sequence sort itself will be called **seq**. If the base signature τ was itself multi-sorted (as it would be if we were studying the logic of programs of a real program language) then there would be a sequence sort for each base sort (of course in the programming example **nat** would most likely already be a base sort). We add the customary constant, function, and predicate symbols to stand for the usual operations over the natural numbers. For finite sequences we allow the operation $s(t)$ where s is a formal variable over **seq** and t is a term of sort **nat**. The resulting term is, of course, of sort **k**.

It is easy to see that the explicit formula (3) is in this extended language if the original χ is. The language is called second-order because of the presence of functions (the sequences); it is weak second-order because the sequences are all finite. We can extend the notion of a "finitary" formula to this language by saying that $\chi(x, A)$ is finitary in A if A occurs positively and only within the scope of essentially existential quantifies (over **k**, **nat**, and **seq**) and bounded universal quantifiers over **nat**. With this restriction χ will define a finitary operator.

5 Inductive Semantics of Flowcharts

We will now apply the above analysis to the correctness formula Φ. We shall see that Φ is inductively defined using a finitary defining condition χ. Actually Φ and other predicates are simultaneously inductively defined using finitary χ_j. The definitions are schematic in that the χ_j are constructed from the uninterpreted flowchart and have the same syntactic form in each interpretation.

5.1 The Correctness Predicate

For simplicity we will assume there is just one program variable and no non-deterministic assignment. As we have seen the latter complication only adds existential quantifiers to path expressions.

Given a flowchart $\mathcal{F}$ chose a set of nodes other than the START and STOP nodes such that all loops pass through one of the chosen nodes. These nodes are called"cut nodes". For the present discussion this choice is arbitrary but when we later consider structured languages we will see that imposing structure on programs determines appropriate "cut nodes" thus simplifying the analysis. To each such cut node, η, attach a new predicate symbol $Q_\eta(y, w)$. If the cut node is an ACTION node the predicate symbol is attached after the node (i.e., before all out arrows). We are going to define Φ by giving a simultaneous inductive definition of Φ and all the Q_η.

Over a given $\mathcal{A}$ the intended meaning of Q_η is:

$$\mathcal{A} \models Q_\eta(a, b)$$

$$\Leftrightarrow$$

There is a run of $\mathcal{F}(a)$ with $x_{current}$ equal to b when control reaches η.

Note that this interpretation makes the Q very much like Φ; indeed we can think of Φ as the disjunction of Q's attached to STOP nodes.

Consider all paths P through $\mathcal{F}$ that begin either at START or some cut node and end either at a STOP or a cut node and don't pass through any intermediate cut node. There are finitely many such P. Each corresponds to a straight-line program from its beginning to end (which might be the same node). For each such path P we construct a formula. There are four kinds of P

1. P begins at START and ends at a cut node η
2. P begins at START and ends at a STOP node
3. P begins at η_1 and ends at η_2
4. P begins at η_1 and ends at a STOP node.

The first two kinds of paths correspond exactly to straightline programs. We begin with y and follow the path P to η or a STOP node and get a formula of the form

$$\phi_P(y) \wedge w = t_P$$

as we've seen before. The first part says that the path P has been followed and the second part symbolically executes along that path to yield a term t_P, formed using the variable y which represents the value in x when η (or the STOP node) is reached. We call the above formula $\chi_P(y, w)$; note that none of the Q predicates appear in it.

Given a path P of the third or fourth kind from η_1 to η_2 or to a STOP node we now formulate its χ_P. Unlike paths of the first two kinds which begin at START we don't know the value of the program variables when we begin at η_1. Here's where inductive definability comes in; we just assume we do know the values! This is done by letting $\chi_P(y, w)$ be:

$$\exists v[Q_{\eta_1}(y, v) \land \phi_P(v) \land w = t_P]$$

Here $\phi_P(v)$ and t_P are as before but v is used as the starting values in x at η_1 (so that $\phi_P(v)$ is a condition on v and the term t_P has v as its only variable). Note that only the Q corresponding to the beginning node of P occurs in χ_P.

Finally, we consider a simultaneous inductive where there is a clause for each path P as defined above. If P ends in η_2 then there is a clause

$$\chi_P(y, w) \Rightarrow Q_{\eta_2}(y, w)$$

and if P ends in a STOP node then there is a clause

$$\chi_P(y, w) \Rightarrow \Phi(y, w).$$

We summarize the above in a theorem.

Theorem 7 *The correctness predicate Φ for non-deterministic flowcharts is schematically inductively definable using a schematic finitary defining condition. The form of the definition depends on a choice of cut points. The resulting Φ is weak second order.*

5.2 Proofs of Correctness

We now reach the completion of this report. The full correctness relation Φ is usually not the main focus of interest. One is usually given a program and a statement $\Psi(x, y)$ and wish to show

$$\Phi(y, z) \Rightarrow \Psi(y, z)$$

which means that if the program with input y has a run which halts and z is the final value in the variable then $\Psi(y, z)$. This is the condition called "partial correctness" since there is no claim that the program halts on any run. Combining all the work we have done previously we see that this result can be proved if we can find predicates Q^*_η which satisfy the verification conditions which result from the form of the inductive definition. If the correctness condition is true then such predicates always exist and can be taken to be in weak second order form (namely the real Q_η defined by the inductive definition). The real interest arises when Ψ is first order and the Q^*_η can also be taken to be first order. While this is not always the case it is when the underlying data type is rich enough (namely rich enough to encode finite sequences) as it is in programs over the integers. We return to the "sufficiently rich" assumption in later parts of this work.

The fact that proofs of the verification conditions suffice to show program correctness was first point out with a less general proof by Floyd.

Prolog Programming

Gerald E. Sacks

I. Introduction to the Prolog Language

Good morning. In these five lectures I hope to show you some of the possibilities of prolog programming.[1] The first two will cover the elements of the prolog language, the third will discuss the CNV inference engine, the fourth will be on definite clause grammars, and the fifth on priority prolog.

The best preparation for prolog programming is classical recursion theory, particularly the dynamic aspects initiated by Post in his simple set construction. Prolog is a higher level language than *C* or pascal, but the prolog programmer reaps considerable benefits from keeping in mind how prolog is implemented. The execution of a prolog program resembles a tree traversal. The dynamics of execution cannot be ignored by the prolog programmer.

The language is small. Keep in mind that it is case sensitive. A useful slogan at this point: everything is a structure.

A *variable* is any sequence of alphanumeric characters that begins with an uppercase letter or an underscore. In addition, a single underscore indicates the so-called anonymous variable, which behaves somewhat like an existentially quantified variable.

An *atom* is a constant, a sequence of ascii symbols. Usually an atom begins with a lower case letter. If it begins with an upper case letter or a numeral, then it must be enclosed in single quotes. Examples of atoms are:

herbert   'Herbert'   '4a'   &   .

---

[1] The first four lectures of this short course on prolog are based on "The Arity/Prolog Language Reference Manual" (Arity Corporation, Concord, MA 1988) and "Prolog Programming In Depth" by Covington, Nute and Vellino (Scott Foresman, 1988). These two excellent sources are recommended for further study.

Integers and floating-point numbers are also available. Some examples of the latter are:

$$1.60 \qquad -0.59 \qquad 147.3e25 \ .$$

None of the following are floating point numbers:

$$77. \qquad .63 \qquad -.47 \ .$$

All of the above are atomic structures. Complex structures are comprised of functors and simpler structures. In place of a recursive definition I offer some examples.

$$\text{person(Name, hair(Color, Length))} \tag{1}$$

person and hair are functors. Any atom can serve as a functor. Name, Color and Length are variables. person is said to have an arity of 2 (i.e., the number of arguments), more briefly person/2. Similarly hair has arity 2. The structure, hair (Color, Length), has been substituted for the second argument of person. Now consider

$$\text{person(herbert, hair(brown, Length)).} \tag{2}$$

Name in (1) has been instantiated to herbert, and Color to brown. We could have started with person(X, Y), and then obtained (1) by instantiating X to Name, and Y to hair (Color, Length).

Usually, structures are represented in prefix Polish notation, but prolog permits operators to be defined, or redefined, in infix notation.

The fundamental mechanism of prolog is *unification*. It may or may not succeed.

(a) Any variable can be unified with any other variable. After unification, if one becomes instantiated, then so does the other to the same value.

(b) A variable can be unified with a structure. For example, X with herbert, or Y with (1). This case is similar to the role of assignments in C or Pascal.

(c) Two integers, if equal, can be unified. The same holds for atoms. Two floating point numbers can never be unified.

(d) Two complex structures can unify, if their functors match (same name and arity), and if their arguments unify.

For the sake of speed, implementations of prolog leave out the so-called "occurs check". It prevents unification of X with $f(X)$. Thus in real life, circular unifications are allowed, and the result is usually a crash.

Lists are available in Prolog. For example

$$\text{[tom, dick, harry]} \tag{3}$$

is a list with three elements. A list has a head and a tail. The head of (3) is the element tom, and the tail is the list [dick, harry]. A vertical bar | can be used to separate them. Thus (3) is equivalent to

$$\text{[tom | [dick, harry]].} \tag{4}$$

The unification of [H| T] with (3) will succeed. H unifies with tom, and T with [dick, harry]. A handy feature of lists in prolog, not common to Lisp, is an arbitrary head. For example,

[herbert, tom | T]

is the same as

[herbert, tom, dick, harry].

Let's look at some prolog code.

```
(5.1)      parent(X,Y) :-
               father(X,Y) .

(5.2)      parent(X,Y) :-
               mother(X,Y) .

(5.3)      father(jim, sam).
(5.4)      mother(betty, sam).
(5.5)      mother(luisa, ella).
```

(5.1) and (5.2) are rules. The intention of (5.1) is: X is a parent of Y if X is the father of Y. (5.3) and (5.4) are facts. (5.3) says jim is the father of sam. Suppose the above code is loaded, and we ask:

(6) ? parent(X, sam).

The interpreter begins at the top, looks for a match with parent, and finds it with (5.1). Then X is unified with X, and Y with sam. Now back to the top with father (X, sam) in search for a match with father, found with (5.3). X is unified with jim, and the interpreter answers the original query with: X = jim.

The query ? parent(X, herbert) will be answered with: no.

Query(6) has a second solution; X = betty, which can be obtained by pressing ";" after the first solution is reported.

(5.1) and (5.2) can be combined into one rule:

```
parent(X,Y) : -
        father(X,Y) | mother(X,Y).
```

The vertical bar is like "or". The goal parent (X,Y) succeeds if father (X,Y) succeeds, or if mother (X,Y) succeeds.

Comma is used for conjunction.

```
a: - b,c .
```

means a if (b and c).

Call (P) succeeds iff P succeeds. Note that the variable P can range over structures. In prolog scant attention is paid to type differences. A variable ranges simultaneously over integers, floating point numbers, atoms, complex structures, etc. This feature of prolog, sometimes termed *metavariables*, is extremely powerful.

Let's examine the query: ? parent(X, herbert) in more detail. The first match is with (5.1). Y is unified with herbert. The goal parent(X, herbert) is replaced by the subgoal father(X, herbert). The next match is with (5.3). X unifies with jim, but herbert does not unify with sam. father(X, herbert) has failed. parent(X, herbert) has failed. Now the phenomenon of backtracking occurs. The initial match with (5.1) failed, so a match with (5.2) is tried. It too fails in the end.

In general when a goal fails, prolog will try an alternative. The backtracking mechanism undoes all the unifications that took place during the execution of the first alternative before it attempts the second. If ? parents (X, herbert) is traced, then eventually mother (X, herbert) is tried. The first match is with (5.4), which instantiates X to betty. It fails and backtracks to (5.5), but first instantiates to X, and then instantiates to luisa.

The cut symbol, !, can be used to control backtracking. Consider:

(7.1) a : - b,c,d .

(7.2) c : - e, f, !, g.

Suppose ? a. Assume b succeeds. Now c is the goal. Assume e and f succeed. The cut symbol, !, is a goal that always succeeds. Assume g succeeds. Thus c succeeds. Suppose d fails. Then prolog backtracks to c. The first goal to be retried is g. Suppose all alternatives to g fail. Without the cut, prolog would seek alternatives to f. With the cut, no alternative to f is sought, and no alternative to c is sought; instead the next move is to seek an alternative to b.

Thus a cuts out subgoals on its left, and its parent as well. Cuts have many uses, but are dangerous. A form of negation can be defined using cut.

(8.1) not(G) : - call(G), !, fail.

(8.2) not(G) .

fail is a goal that always fails. Consider

(9) not(not(P(X))).

(9) succeeds if and only if P(X) has a solution. The utility of (9) is the fact that ? (9) will not instantiate X when (9) succeeds.

All cuts can be eliminated with the use of not. This is not always a good idea.

A final word for today: P(-,Y) succeeds iff P(X,Y) succeeds, but the query ? P(-,Y) reports only the solution for Y. The anonymous variable _ in place of X seems to have the same effect as an existential quantifier on X.

2. Recursion and other Prolog Techniques

Let us compare recursion in prolog with recursion in Lisp. Covington, Nute and Vellino (Scott, Foresman 1988) say they are quite different, because of prolog's use of uninstantiated variables. Recall append, a standard procedure for concatenating lists. It returns [a,b,c,d,e] when its inputs are [a,b] and [c,d,e]. In prolog,

```
append([ ], X, X] .
append(Head | Tail], X, [Head | Y]) : -
   append(Tail, X, Y).
```

In Lisp,

```
(DEFUN APPEND (LIST1 LIST2)
    (IF ( NULL LIST1)
        LIST2
        (CONS (CAR LIST1) (APPEND (CDR LIST1)
        LIST2))))
```

The way it works in Lisp is: If list1 is empty, return list2; otherwise return a list whose first element is that of list1, and whose tail is obtained by appending list2 to the tail of list1. Note the cons operation is performed after the recursive call to append. Thus the Lisp version of append is not tail recursive. (The latter means that the recursive call is not the last step.)

Here is the way it works in prolog: split list1 into Head and Tail; create a third list whose head is Head, and whose tail is Y, an uninstantiated variable. Of course Y merely points to a place in memory where the tail will go. The Prolog version of append is tail recursive thanks to Y.

We turn now to a more detailed discussion of terms. Some of the data classification predicates available in Prolog are:

atom(X), succeeds if X is an atom;

integer(X), succeeds if X is an integer;

var(X), succeeds if X is an uninstantiated variable.

A good example of cut and fail is the definition of nonvar.

```
nonvar(X) : -
        var(X), !, fail.
nonvar(_) .
```

The unification predicate, X = Y, succeeds if X and Y unify. X = 1 succeeds if X is uninstantiated or if X is already unified with 1. X \= Y succeeds if X cannot be unified with Y. The comparison predicate, X == Y, succeeds if X and Y, after evaluation, are equivalent. X == X always succeeds. eq(X,Y) succeeds if X and Y are the same structure and occupy the same memory location.

The so-called univ predicate, = ..., converts structures to lists and lists to structures

```
? - book(life, vander) =. . X.
X = [book, life, vander].
```

```
? - X = . . [book, life, vander].
X = book(life, vander).
```

Univ can be used to add arguments to a structure.

```
addargs(Struct, List-of-args, Struct2) : -
        Struct =. . List,
        append(List, List-of-args, List2),
        Struct2 =. . List2.
```

A related predicate is functor (Structure, Name, Arity). It returns the name and arity of a structure. Another is arg(Arg, Struc, Value). It returns the value of an argument in a structure.

```
? -   arg(1, book(life, vander), V).
V = life.
```

Terms can be collected with bagof(T,P, List).

```
is(herb, happy).
is(jim, sad).
is(low, tall).
is(fred, happy).
is(fred, tall).

? - bagof(Y, is(Y, happy), L).
L = [herb, fred].

? - bagof(X, Y^ is (Y,X), L).
L = [happy, sad, tall, happy tall].
```

The effect of Y^ is the same as ∃ Y.

Let's return to recursion. An old favorite is factorial.

```
factorial(O, M, M).
factorial(N, M, Ans) : -
        N1 is N-1,
        M1 is N * M,
        factorial(N1, M1, Ans).
```

The above is tail recursive. A decent implementation of prolog transforms tail recursion into iteration.

Note: a failed unification such as factorial (O,M,M) with factorial(N,M,Y) is faster than N > 0.

The precise definition of tail recursive is: procedure P is tail recursive iff P calls itself with an empty alternative set and an empty continuation.

The alternative set consists of other clauses for P below the recursive call to P. The continuation set consists of subgoals after the recursive call to P.

If both sets are empty, the recursive call to P places no additional information on the stack.

More on cuts.

Green cuts make procedures more efficient. If a green cut is removed, the program still works correctly. Example:

```
max(X,Y,X) :- X >= Y, ! .
max(X,Y,Y) :- Y > X .
```

Now remove the cut

```
max(X,Y,X) :- X >= Y .
max(X,Y,Y) :- Y > X .
```

Now consider

```
max(X,Y,X) :- X >= Y, ! .
max(X,Y,Y).
```

If the last cut above is removed, the program produces incorrect results when $X \geq Y$ and backtracking occurs.

Red cuts alter the declarative meaning of a program. They cannot be removed. The cut fail combination is an example of a red cut.

```
nonvar(X) :-
        var(X), !, fail.
nonvar(_).
```

The effect of a cut inside a recursion can be perplexing. The following example is obtained from some Arity Prolog documentation.

```
vary([ ],[ ]) :- ! .
vary([H],[H]) :- ! .
vary([H | T, [H | T2]) :-
    vary(T, T2).
```

```
vary([A,B | T], [B, A | T2]) :-
        vary(T, T2).
```

The first solution obtained for X, in vary([a,b,c], X) is [a,b,c]. The second solution is [a,c,b] as a result of backtracking from the third clause. The point to remember is: a cut is valid only at its recursion level. The third solution is [b,c,a].

For the next lecture some information about operators is needed.

The predicate op(Prec, Assoc, Op) can be used to define the precedence and associativity of an operator. Precedence has a scale of 1 to 1200. Lower precedence means: do it first. Two standard definitions are:

```
op(500, yfx, +).
op(400, yfx, *).
```

yfx means the operator is infix and left associative. Thus a*b + c is read as +(*(a,b),c). The complete table for f is:

infix	xfx	nonassociative
	xfy	right to left
	yfx	left to right
prefix	fx	nonassociative
	fy	left to right
postfix	xf	nonassociative
	yf	right to left

The CNV inference engine uses:

```
op(900, fx, neg).
```

Consider:

```
neg blah(a,b) :-
        foo(a,b).
```

A standard definition is: op(1200, xfx, :-). So the above use of neg is read as

```
(neg blah(a,b)) :-
        foo(a,b).
```

This is of course the way one would expect neg to behave. Prolog actually reads the above as

```
:- (neg(blah(a,b)), foo(a,b)).
```

In Prolog there are only structures; : - is just another functor. As an operator, : - is infix and nonassociative. A last word on operators: parentheses override operator definitions. A last last word: spaces between operators and arguments often matter a great deal; a space is needed between neg and blah; it may or may not be needed between + and 2.

3. The CNV Inference Engine

Covington, Nute and Vellino (Scott, Foresman 1988) present an inference engine based on negation and defeasible rules. Their code is an excellent example of the power of prolog. Their engine is the most interesting I have seen.

A defeasible rule is one that admits exceptions. It is written with a new operator : =. Thus

flies(X) : = bird(X).

means: X flies if X is a bird and X is not an exception. The operator neg will be needed to indicate exceptions, as in

neg flies(X) : = penguin(X).

The inference engine uses @ to query goals, as in

? - @ Goal.

The above query is successful if Goal is successful in the sense of the engine. The code for the CNV engine is as follows.

```
init : -  op(1100, fx, @),
          op(900, fx, neg),
          op(1100, xfx, : =),
          op(1100, xfx, : ^).

: - init.

@ Condition : -    Condition =.. [',', First, Rest],
                   ! ,
                   @ First ,
                   @ Rest .
```

```
@ Goal : - Goal.

@ Goal : - clause (Goal, Condition) ,                              (1)
           Condition \ = true ,
           @ Condition,
           opposite(Goal, Contrary),
           not contrary.

@ Goal : - (Goal : = Condition),                                   (2)
           @ Condition
           opposite(Goal, Contrary),
           not Contrary,
           not defeat (Goal : = Condition).

opposite(neg Clause, Clause) : - ! .
opposite(Clause, neg Clause).

defeat((Head := Body)) : -                                         (3)
           opposite(Head, ContraryOfHead),
           clause(ContraryOfHead, Condition),
           @ Condition.

defeat((Head : = Body)) : -                                        (4)
           opposite(Head, ContraryOfHead),
           (ContraryOfHead : = Condition),
           not_ more_ informative(Body, Condtion),
           @ Condition.

defeat((Head : = Body)) : -                                        (5)
           opposite(Head, ContraryOfHead),
           (ContraryOfHead :^ Condition),
           not_ more_ informative(Body, Condition),
           @ Condition.
```

```
    not_more_informative(Clauses1, Clauses2) :-
            not absolute_consequence(Clauses2, Clauses1).

not_more_informative(Clauses1, Clauses2) :-
            absolute_consequence(Clauses1, Clauses2).

absolute_consequence(Goals, Premises) :-
            Goals =.. [',', first, Rest],
            absolute_consequence(First, Premises),
            absolute_consequence(Rest, Premises).

absolute_consequence(true,_).

absolute_consequence(Goal, Premises) :-
            belongs(Goal, Premises).

absolute_conseqeunce(Goal, Premises) :-
            clause(Goal, Body),
            Body \= true,
            absolute_consequence(Body, Premises).

belongs(Clause, Clause).

belongs(Clause, Conjunction) :-
            Conjunction =.. [',', Clause,_].

belongs(Clause, Conjunction) :-
            Conjunction =.. [',',_, RestOfConjunction],
            belongs(Clause, RestOfConjunction).
```

The first portion of code creates four operators. @ is a query operator appropriate for the code. Let us see how @(Goal) works. It will certainly succeed if Goal succeeds in the normal sense of prolog. It will succeed if there is a prolog rule

Goal :- Condition,

and all the clauses in Condition succeeds in the sense of @. Note how =.. is used to define @ for a conjunction of clauses (i.e., a condition). Clause(G,C) works as follows. Suppose a : - b is in the data base, and the query ?-clause(a,X) is made. Then the reply is X = b.

Block (1) of code begins by looking for an ordinary prolog rule to derive Goal. A check is then made to make sure Condition is not a fact in order to eliminate some duplicate solutions. Next an attempt on deriving Condition in the sense of CNV. Finally a check to make sure that Goal is not outright false.

Block (2) of code makes use of : =. The CNV engine introduces defeasible rules. a : = b is a rule with exceptions; it can be defeated . The notion of defeat begins with block (3). a : = b can be defeated if neg a can be deduced in the sense of @. It can be defeated (block (4)) if neg a follows from some condition via a defeasible rule, if a is not more informative than the condition, and if the condition is derivable in the sense of @.

Block (5) is similar to block (4). Now neg a has to follow from a (so-called) defeater,

neg a :^ Condition.

The operator :^indicates the presence of a defeater. A defeater is a rule whose sole use is to defeat a defeasible rule; it never leads to any conclusions.

The notion of absolute_ consequences figures in not_ more_ informative. It is the usual notion of consequence restricted to ordinary prolog rules.

Covington *et al* consider the following example.

i) Normally, birds fly.
ii) Penguins never fly.
iii) If something is sick, then it might not fly.
iv) Presumably Buzz flies.
v) Woody is a bird.
vi) All penguins are birds.
vii) Chilly is a penguin.

The above translateds to:

i) flies(X) : = bird(X).
ii) neg flies(X) : - penguin(X).
iii) neg flies(X) :^ sick(X).

iv) flies(buzz) : = true.

v) bird(woody).

vi) bird(X) : - penguin(X).

vii) penguin(chilly).

With the above in the database, the query @ flies(X) yields X = woody and X = buzz. The query @ neg flies (X) yields X = chilly. Now add

penguin(woody).
sick(buzz).

to the database. Now the query @ flies(X) yields no solution, and the query @ neg flies(X) yields X = woody and X = chilly. Note that @ neg flies(X) does not yield X = buzz.

Here is a much more complicated example from Covington *et al.*

Presumably, the free traders will nominate Hunter.

If not, then they will nominate Farmer.

They will nominate Baker if they nominate neither Hunter nor Farmer, presumably.

They will not nominate Hunter, if Gardner does not support him, presumably.

The isolationists will nominate Fox if Bull does not run and the free traders do not nominate Hunter, presumably.

But they will nominate Bull if he runs and Crow supports Fox, presumably.

If they do not nominate Fox or Bull, they will nominate Hart, presumably.

The free-trader candidate will be elected if Crow supports the isolationist candidate, presumably.

But Bull will be elected if he gets the isolationist nomination and Gardner does not support the free-trader candidate, presumably.

No one supports two different candidates.

Hunter, Farmer, Baker, Fox and Hart run.

Presumably, Bull does not run.

Gardner supports Baker.

Baker supports Fox.

The query @ nominate(P,C) yields free traders nominate farmer and isolationists nominate fox. Who is elected. (Answer at the end of Section 5.)

4. Definite Clause Grammars

This lecture follows the discussion of DCG's given in "The Arity/Prolog Language Reference Manual" (Arity Corporation, Concord, MA 1988).

First consider context free grammars (CFG). In place of a definition, a carefully chosen example from documentation supplied by the Arity Corporation. Here in ordinary language is a CFG designed to recognize the sentence "the musician plays the violin".

a sentence is a noun phrase followed by a verb phrase.

a noun phrase is a determiner followed by a noun.

a verb phrase is a verb followed by a noun phrase.

a determiner is the.

a noun is musician or violin.

a verb is plays.

The above in Backus-Naur form (BNF) is:

(i) < sentence > : : = < noun_ phrase > < verb_ phrase >

< noun_ phrase > : : = < determiner > < noun >

< verb_ phrase > : : = < verb > < noun_ phrase >

(iv) < determiner > : : = the

< noun > : : = musician | violin

< verb > : : = plays.

The above has a simple interpretation with the help of difference lists. Let S,T,U... be lists of words, possibly empty. S-T is the set-theoretic difference, but is used only when T is a final segment of S. For example, S is [the musician plays the violin] and T is [plays the violin].

Now (i) can be rendered as

S-U is a sentence if there exists a T such that S-T is a noun phrase and T-U is a verb phrase.

And (iv) becomes

S-T is a determiner if S is a list whose head is the and whose tail is T.

All this is easily expressed in prolog.

```
(i)*   sentence(S,U) : -
              noun_ phrase(S,T), verb_ phrase(T,U).
```

```
noun_ phrase(S,U) :-
    determiner(S,T), noun(T,U).

verb_ phrase(S,U) :- verb(S,U).
verb_ phrase(S,U) :-
    verb(S,T), noun_ phrase(T,U).

determiner([the | S], S).

noun([musician | S], S).
noun([violin | S], S).

verb([plays | S], S).
```

The query ? sentence ([the musician plays the violin], [ ]).
yields yes.

Despite the limitations of CFG's, they are powerful enough to specify programming languages such as Pascal and C. What's missing is number, the ability to recognize the musicians play as well as the musician plays. From a procedural view, what's missing is the ability to pass parameters. The prolog language uses the symbol - -> to render BNF's. Thus (i) becomes

```
(i)**    sentence - - >
             noun_ phrase, verb_ phrase.
```

Of course a prolog interpreter reads (i)** as if it were (i)*. To pass a parameter, alter (i)** to

```
(i)***   sentence - - >
             noun_ phrase (Number), verb (Number).
```

The above is an instance of a DCG (definite clause grammar).
The interpreter reads (i)*** as

```
sentence (S,U) :-
    noun_ phrase (N, S, T),
    verb_ phrase (N, T, U).
```

The ability to pass parameters is all that is needed to compute anything that's computable. Thus DCG's are universal. Standard prolog allows a mixing of prolog notation and DCG notation.

A famous example of DCG notation is due to Pereira and Warren, Journal of Artificial Intelligence, [1978]. They wrote a DCG that parses a sentence and builds a structure that analyzes the sentence.

```
: - op(910, xfy, & ).
: - op(920, xfy, =>).
: - op(930, xfy, : ).

sentence(P) - - >
    noun_ phrase (X, P1, P),
    verb_ phrase (X, P1).

noun_ phrase(X, P1, P) - - >
    determiner(X, P2, P1, P),
    noun(X, P3),
    rel_ clause(X, P3, P2).
noun_ phrase(X, P, P) - - > name (X) .

verb_ phrase(X,P) - - >
    trans_ verb(X, Y, P1),
    noun_ phrase(Y, P1, P).
verb_ phrase(X,P) - - >
    intrans_ verb(X,P).

rel_ clause(X, P1, P2) - - >
    [that],
    verb_ phrase(X, P2).
rel_ clause(_ ,P,P) - - > [ ].
```

```
determiner(X, P1, P2, all(X) : P1 => P2) - - >
     [every].
determiner(X, P1, P2, exists(X) : P1 & P2) - - >
     [a].

noun(X, man(X)) - - - > [man].
noun(X, woman(X)) - - > [woman].

name(john) - - > [john].
name(mary) - - > [mary].

trans_ very(X,Y, loves(X,Y)) - - >
     [loves].
intrans_ verb(X, lives(X)) - - >
     [lives].
```

The last line of the above, rendered in prolog, is:

```
intrans_ verb(X, lives(X), [lives | B], B).
```

Suppose the above program is queried with

```
sentence(X, [every, man, that, lives,
     loves, a, woman], [ ]).
```

The response is X =

```
all(_ 0088) : lives (_ 0088) =>
     (exists(_ 01AC) : woman(_ 01AC)
          & loves(_ 0088, _ 01AC))
```

(This last is output from Arity version 5. _ 0088 and _ 01AC are memory locations of uninstantiated (i.e., free) variables.)

5. Prolog and Priority

Imagine a goal that consists of a huge number of subgoals. Assume that many of the subgoals conflict with each other; i.e., they have no common solution. The priority method of recursion theory can now be applied.

For the sake of simplicity, consider a special situation with the above properties. A huge database is available. A typical fact from the database is of the form:

job J_i can be done by team T.

A team T consists of two parts: T^+, the positive, and T^-, the negative. T^+ is a set of persons who can work together to do job J_i so long as no person from T^- is present. Thus the choice of T to do J_i means hiring T^+ and excluding T^-. More abstractly, T is an ordered pair $< T^+, T^- >$ of disjoint sets.

The notion of extension is (perhaps) unorthodox. Let T_a and T_b be teams. T_a is *extended* by T_b iff $T_a^+ \subseteq T_b^+$. This notion of extension allows $T_b^+ \cap T_a^-$ to be non-empty. Hence it can happen that T_a can do J_i, T_a is extended by T_b, but T_b cannot do J_i. This is the appropriate notion of extension for the priority construction of a team that can do many jobs simultaneously.

The extension rule is: if T_a can do J_i, T_a is extended by T_b and $T_b^+ \cap T_a^- = \phi$, then T_b can do J_i. This rule makes it possible to construct a team that can do many jobs simultaneously.

T_a is said to *rule out* T_b if $T_a^+ \cap T_b^- \neq \phi$. Suppose T_a can do J_i and $T_a^+ \subseteq T_b^+$. Suppose T_a is chosen. If T_a rules out T_b, then it would be senseless to extend T_a to T_b. In dynamic terms, once T_a is chosen it is too late to choose T_b.

$\lambda_n | T_n$ is said to be a sequence of teams if T_n is extended by T_{n+1} for all n. (n ranges over a large, finite, initial segment of the natural numbers.) T_n does J_i forever iff T_n does J_i and $T_m^+ \cap T_n^- = \phi$ for all $m > n$.

d is a designation function for $\lambda_n | T_n$ iff for all i: if for some n T_n does J_i forever, then $T_{d(i)}$ does J_i forever. Note that d is a partial function.

The problem addressed by Prolog with Priority (in this special situation) is: find a $\lambda_n | T_n$ and a d such that for all i: either (a) or (b) holds.

(a) There is an n such that T_n does job J_i forever.

(b) For all T_a and i, if "T_a does J_i" is a fact in the database, then either (b1) or (b2) holds.

(b1) There is an n such that T_n rules out T_a.

(b2) There is a $j < i$ such that $T_a^+ \cap T_{d(j)}^-$ is not empty.

If $\lambda_n|T_n$ and d satisfy (a) or (b) for all i, then d is said to be a satisfactory designation function for $\lambda_n|T_n$.

J_i is said to have higher priority than J_k if $i < k$.

Clause (b2) says there is a j such that J_j has higher priority than J_i and $T_a^+ \cap T_{d(j)}^-$ is not empty.

What is the dynamic meaning of clauses (a) and (b)? The best possible outcome for d would be (a) holding for all i. Fix i and suppose (a) fails. Suppose further that "T does J_i" belongs to the database. Then (b) explains the failure of $\lambda_n|T_n$ to incorporate T. If (b1), then T_n was chosen before "T does J_i" became available, and $T_n^+ \cap T^- \neq \phi$; thus "T does J_i" came along when it was too late to use it. If (b2), then there is a $j < i$ such that $T_{d(j)}$ was chosen before "T does J_i" because available, and $T^+ \cap T_{d(j)}^- \neq \phi$; thus "$T$ does J_i" came along at a time when the use of it would undo a choice made earlier on behalf of a job of higher priority. In short (b) says that every opportunity to do J_i appeared too late, or at the wrong time with respect to priorities.

(b2) is clearer after examining a proof of Theorem A (not given in these notes but sketched in these lectures). Assume a database; then

THEOREM A.. *There exists a sequence $\lambda_n|T_n$ of teams with a satisfactory designation function d. In addition $\lambda_n|T_n$ and d are developed in one pass, with intelligent backtracking, through the database.*

Most of the power of Theorem A is contained in its second sentence. Unfortunately it is not possible to define intelligent backtracking without proving the first part of Theorem A. Suffice it to say the procedure for developing $\lambda_n|T_n$ and d is the finite injury method of Friedberg and Muchnik.

It turns out there is a bound of n^2 on the steps taken by the backtracking mechanism. The usual exponential blowup is avoided at the price of not satisfying subgoals of low priority. The proof of Theorem A is what R. Soare calls a $O^{(1)}$ construction. It is possible to develop results similar to Theorem A for $O^{(k)}$ arguments for all $k \geq 1$. Then

STATEMENT B:. *If a prolog-with-priority construction is of type* $O^{(k)}$, *then* n^{k+1} *is a bound on the number of steps taken during backtracking.*

In order to state Theorem C, a continuity result, a few last definitions are needed. The database D is a set of facts $\{F_z | z < n\}$. If $z_1 < z_2$, then D_{z_1} precedes F_{z_2}. Suppose D is truncated to E. E might be $\{F_z | z < m\}$ for some $m < n$. Let $s(D)$ be the solution obtained by subjecting D to the priority construction of Theorem A. Define $s(E)$ similarly. Let $s_t(D)$ be the value of $s(D)$ at the beginning of step t of the priority construction. Then

THEOREM C.. *Let* t_E *be the first step at which the priority construction for* D *reads a fact in* $D - E$. *Then*

$$s_{t_E}(D) = s(E).$$

(The answer promised at the end of Section 3 is: farmer is elected.)

A Guide to Polymorphic Types

ANDRE SCEDROV

University of Pennsylvania
Departments of Mathematics and Computer Science

INTRODUCTION

Types have now become an important ingredient of programming language design as a powerful, flexible syntax of a logic of program specifications that can be incorporated into a programming language itself. Types provide both a context for an organized, logical development of programs according to given specifications and a framework for a partial verification mechanism (see e.g. Breazu-Tannen et al. [88]). These features are crucial in large-scale programming efforts that require coordination among many teams of programmers.

One of the most important aspects of recently developed programming languages such as *ML-like languages* (Gordon et al. [79], Milner [84], MacQueen [85], Cousineau [87]), *Ada* (Barnes [81]), *Miranda* (Turner [85]), and *Clu* (Liskov [81]) is the way in which they extend the conventional type systems in the Algol/Pascal family of programming languages: they feature polymorphic or generic data types that allow programmers a new form of flexibility and abstraction in programming.

Among various notions of polymorphism first introduced in Strachey [67], the most influential one is the concept of *parametric* (or: *horizontal*) *polymorphism*. Intuitively, a parametric polymorphic function is one that has a uniformly given algorithm in all types. One of the telling examples is Strachey's map-list example:

EXAMPLE. Consider a function f whose argument is of type p and whose result is of type q, so that the type of f is p⇒q. Let L be a list of elements of type p. We may say that L is of type p list. Now consider the following function map : apply f to the entries in L, then make a list of the results. Thus map f L is a list of elements of type q (application is associated to the left). The function map is of type (p⇒q) ⇒ p list ⇒ q list (in type expressions, association is to the right). Note that we have not referred to any information about p and q. ∎

Another good example of a parametric polymorphic function is the list iterator list_it, used in iterating along lists in reverse order. The function list_it, like map, is an ML primitive.

EXAMPLE. Let h be a function whose argument is of type p and whose result is of type q⇒q. Let [L1; L2; ... ; Ln] be of type p list, let x be of type q. On these data list_it computes (h L1 (h L2 (... (h Ln x)...))) of type q. list_it is of type ((p⇒q⇒q) ⇒ p list ⇒ q ⇒ q). ∎

The study of parametric polymorphism in programming languages has focused in the recent years on various rich type systems in the framework of typed lambda calculi. These type systems allow types to depend on other types and possibly on ordinary values. A formal calculus of such *variable types* was developed in Girard [71,72] as a higher-order extension of the Curry-Howard propositions-as-types paradigm in pure mathematical logic. Second-order fragment of this calculus was independently proposed in Reynolds [74] as a syntax that captures Strachey's notion of parametric polymorphism. Several powerful extensions of the Girard-Reynolds calculus have been studied and implemented, most notably the Coquand-Huet calculus of constructions, see e.g. Huet [87].

In these expository lecture notes we highlight basic facts about syntax and semantics of polymorphic typed lambda calculi. Second-order polymorphic lambda calculus is introduced in section 1. Sections 2 and 3 contain complete proofs of the confluence theorem and the strong normalization theorem for this calculus. An overview of current research in semantics of polymorphism is given in section 4.

In section 5 we introduce a version of the Coquand-Huet calculus of constructions, one of whose interpretations is then discussed in section 6.

We would like to thank Val Breazu-Tannen and Carl Gunter for many stimulating conversations. The author is partially supported by NSF grant CCR-8705596 and by the University of Pennsylvania Natural Sciences Association Young Scientist Award. Section 6 is a modified version of a technical report written in January 1987 for Odyssey Research Associates, Inc. of Ithaca, NY, sponsored by the U.S. Air Force Systems Command, Rome Air Development Center, Griffis AFB, New York 13441-5700, under contract No. F30602-85-C-0098.

1. SECOND-ORDER POLYMORPHIC LAMBDA CALCULUS

Second-order polymorphic types are built inductively from type variables:

$$p \mid A \Rightarrow B \mid \forall p.A \ .$$

Free occurrences of type variables in types are defined as usual in logic. In particular, p is bound in $\forall p.A$ and hence p does not occur free in $\forall p.A$. We use A , B , C , ... to denote second-order polymorphic types, or *types* in short. We identify types that differ only in their bound type variables. $A[B/p]$ is the result of substituting type B for the free occurrences of type variable p in type A , where bound type variables in A may be renamed so as to be distinct from the free type variables in B .

We assume another countably infinite alphabet of ordinary variables, written x , y , z , ... A *context* Γ is a (finite) list of expressions $x{:}A$, where x is an ordinary variable and A is a type, and no x appears twice. An expression $x{:}A$ may be read as " x has type A ". $\Gamma[A/x]$ is obtained from Γ by adding $x{:}A$ and striking out $x{:}B$ if any such appears. *Second-order polymorphic lambda terms* will be the terms given by the following inductive definition of the *typing judgments* "In context Γ term t has type A ", written $\Gamma \vdash t{:}A$. The definition is given by deduction rules for deriving typing judgments. One simultaneously defines the notion of a *free* occurrence of an ordinary variable in a polymorphic lambda term, as well as a free occurrence of a type variable. We assume that the free occurrences are inherited in a lower line of each rule unless a restrictive comment is made. The rules are:

$\Gamma \vdash x{:}A$ *if* $x{:}A$ *appears in* Γ ,

Lambda abstraction

$$\frac{\Gamma[A/x] \;\vdash\; t : B}{\Gamma \;\vdash\; \lambda x{:}A.\,t \;:\; A \Rightarrow B}$$

(then x does not occur free in $\lambda x{:}A.t$ but p does if it occurs free in A),

Application

$$\frac{\Gamma \vdash t : A \Rightarrow B \qquad \Gamma \vdash u{:}A}{\Gamma \vdash tu : B}$$

Type abstraction

$$\frac{\Gamma \vdash t{:}A}{\Gamma \vdash \Lambda p.t : \forall p.A}$$

if p *is not free in any* B *such that* $x{:}B$ *in* Γ *and* x *free in* t (then p does not occur free in $\Lambda p.t$)

Type application

$$\frac{\Gamma \vdash t : \forall p.A}{\Gamma \vdash t\{B\} : A[B/p]}$$

(then q occurs free in $t\{B\}$ if it occurs free in B),

$$\frac{\Gamma \vdash t{:}A}{\Delta \vdash t{:}A}$$

where Δ *is a permutation of* Γ

We write Γ-x for the context obtained from Γ by striking out any $x{:}B$. Note that if $\Gamma \vdash t{:}A$ and x does not occur free in t , then $\Gamma\text{-}x \vdash t{:}A$. Furthermore, if $\Gamma \vdash t{:}A$ and if $\Gamma,\ x{:}B$ is a context, then $\Gamma,\ x{:}B \vdash t{:}A$.

Let t and u be polymorphic lambda terms of type B and A , respectively (in the same context). We write $t[u/x]$ for the result of substituting u for all free occurrences of the variable x of type A in t, where the bound (i.e., not free) variables in t may be renamed if necessary to prevent clashes (as with types, we identify terms up to renaming of bound variables, i.e. up to alpha conversion). Similarly, let $t[A/p]$ be the result of substituting type A for all free occurrences of type variable p in t (such occurrences may come about from term abstraction or from type application). Detailed and precise treatments of issues related to substitution may be found e.g. in Barendregt [84], Huet [87], and de Bruijn [72].

EXAMPLE 1.1. From $x{:}p \vdash x{:}p$ obtain $\vdash \lambda x{:}p.x : p{\Rightarrow}p$ and therefore $\vdash \Lambda p.\lambda x{:}p.x : \forall p.p{\Rightarrow}p$. Here $\Lambda p.\lambda x{:}p.x$ is "the polymorphic identity". For any type A we get $\vdash (\Lambda p.\lambda x{:}p.x)\{A\} : A{\Rightarrow}A$. (On the other hand, substituting A for p in $\lambda x{:}p.x$ yields $\lambda x{:}A.x$ of type $A{\Rightarrow}A$.) Of course, A may be $\forall p.p{\Rightarrow}p$ itself, in which case $\vdash (\Lambda p.\lambda x{:}p.x)\{\forall p.p{\Rightarrow}p\} : (\forall p.p{\Rightarrow}p) \Rightarrow \forall p.p{\Rightarrow}p$ and we may use the application rule to derive $(\Lambda p.\lambda x{:}p.x)\{\forall p.p{\Rightarrow}p\}\Lambda p.\lambda x{:}p.x$ of type $\forall p.p{\Rightarrow}p$. ∎

An excellent way to think about typing judgments is to consider them as explicit notation for deductions in a natural deduction system for second-order minimal propositional logic. Natural deduction systems, like the related but different sequent calculi, go back to Gentzen and have been studied in Prawitz [65], Stenlud [72], and Girard [72]. Perhaps the quickest way to see this correspondence between terms and natural deductions is to think of types as formulas and then compare the rules given above to those obtained from them by erasing all terms and colons. The latter rules are the rules in the Gentzen-style natural deduction system for second-order minimal propositional calculus. The corresponding typing judgments are in turn obtained by systematic naming of deductions by terms. (Prawitz presents deductions as trees, see example 1.2 below.) Thus in a derived typing judgment $\Gamma \vdash t{:}A$, t names a deduction of A in which the assumptions are exactly the formulas that appear as types in Γ . This correspondence is known as Curry-Howard propositions-as-types paradigm. Curry formulated it in the case of first order minimal propositional calculus (where the terms as those of simply typed lambda calculus) and Howard [69] studied it for first-order intuitionistic arithmetic.

EXAMPLE 1.2. Let p, q, and r be type variables and let Γ be the context $x : r\Rightarrow p\Rightarrow q$, $y : r\Rightarrow p$, $z : r$. Then $\Gamma \vdash z{:}r$ and $\Gamma \vdash x : r\Rightarrow p\Rightarrow q$, whence by application $\Gamma \vdash xz : p\Rightarrow q$. Similarly $\Gamma \vdash yz : p$. Now a use of application rule on these two judgments yields $\Gamma \vdash xz(yz) : q$. Several lambda abstractions now yield that term $\lambda x{:}r\Rightarrow p\Rightarrow q.\ \lambda y{:}r\Rightarrow p.\ \lambda z{:}r.\ xz(yz)$ is of type $(r\Rightarrow p\Rightarrow q) \Rightarrow (r\Rightarrow p) \Rightarrow r\Rightarrow q$ in the empty context. Now type variables may be abstracted in any order. Notice how term application corresponds to the rule Modus Ponens in logic. On the other hand, the term obtained by lambda abstractions is a polymorphic typed version of the **S** combinator. It is in fact an ML program whose specification is the type given above (as type abstractions were not used afterwards in any other rules).

Let us write out the corresponding Prawitz-style natural deduction tree. The applications used at the beginning correspond to

$$\dfrac{\dfrac{r\Rightarrow p\Rightarrow q \qquad r}{p\Rightarrow q} \qquad \dfrac{r\Rightarrow p \qquad r}{p}}{q} \quad ,$$

while the lambda abstraction λz corresponds to the following implication introduction rule which "closes" all occurrences of the assumption r

$$\begin{array}{c} \dfrac{\dfrac{r\Rightarrow p\Rightarrow q \qquad r}{p\Rightarrow q} \qquad \dfrac{r\Rightarrow p \qquad r}{p}}{q} \\ r\Rightarrow q \end{array} \quad ,$$

then two more lambda abstractions yield

$$\dfrac{\dfrac{\dfrac{\dfrac{\dfrac{r\Rightarrow p\Rightarrow q \qquad r}{p\Rightarrow q} \qquad \dfrac{r\Rightarrow p \qquad r}{p}}{q}}{r\Rightarrow q}}{(r\Rightarrow p) \Rightarrow r\Rightarrow q}}{(r\Rightarrow p\Rightarrow q)\ (r\Rightarrow p) \Rightarrow r\Rightarrow q} \quad ,$$

and we could finish with three type abstractions, i.e. quantifier introductions. ∎

EXAMPLE 1.3. Given a set A, the set of lists of elements of A is a generic example of a set B with a distinguished element and a distinguished binary operation which produces an element of B from an element of A and an element of B. In polymorphic lambda calculus, given a type A, the type A *list* is therefore defined as $\forall p.((A\Rightarrow p\Rightarrow p)\Rightarrow p\Rightarrow p)$. In fact, this definition does not assume anything about A, so we may define a generic notion of a list as the type $\forall r.\forall p.((r\Rightarrow p\Rightarrow p)\Rightarrow p\Rightarrow p)$, which we shall call *List*. The empty list is doubly polymorphic and it may be defined as the polymorphic lambda term $\Lambda r.\Lambda p.\lambda y{:}r\Rightarrow p\Rightarrow p.\lambda z{:}p.z$ of type *List* (in the empty context). The polymorphic list constructor that appends an element to a list may be defined as the second-order polymorphic lambda term

$$\Lambda r.\lambda x{:}r.\lambda u{:}(r\,list).\Lambda p.\lambda y{:}r\Rightarrow p\Rightarrow p.\lambda z{:}p.(yx(u\{p\}yz))$$

of type $\forall r.r\Rightarrow(r\,list)\Rightarrow(r\,list)$. ∎

EXERCISE. Express the operations map and list_it discussed in the examples in the introduction as polymorphic lambda terms. ∎

EXAMPLE 1.4. The set of natural numbers is the set of lists on a one-element set, i.e. a generic example of a set B with a distinguished element and a distinguished function from B to itself. In polymorphic lambda calculus we define the type of natural numbers *Nat* as $\forall p.((p\Rightarrow p)\Rightarrow p\Rightarrow p)$. The numeral 0 is defined as the term $\Lambda p.\lambda y{:}p\Rightarrow p.\lambda z{:}p.z$ and the successor S on the natural numbers as $\lambda n{:}Nat.\Lambda p.\lambda y{:}p\Rightarrow p.\lambda z{:}p.(y(n\{p\}yz))$. The numeral m is defined as the polymorphic m-fold iterator, e.g., 2 is represented as $\Lambda p.\lambda y{:}p\Rightarrow p.\lambda z{:}p.y(yz)))$. This representation goes back to Church in the case of untyped lambda calculus. In Girard [72] it is proved that the terms of type $Nat \Rightarrow Nat$ without free variables describe exactly those recursive functions that are provably total in second-order arithmetic (see also Statman [81], Leivant [83]). ∎

A polymorphic lambda term u is a *subterm* of a polymorphic lambda term t if u appears as a contiguous syntactic expression in t which is not prefixed by the symbol λ . Thus u is a subterm of

$\lambda x{:}A.u$, but x is a subterm of $\lambda x{:}A.u$ iff it is a subterm of u. Any polymorphic lambda term is a subterm of itself.

The main computational mechanism on polymorphic terms is the reduction of terms. For any given terms of the appropriate types we define:

$(\lambda x{:}A.t)u$ is *immediately reducible* to $t[u/x]$,

$(\Lambda p.w)\{C\}$ is *immediately reducible* to $w[C/p]$,

and let *reduction* be the least reflexive transitive relation containing immediate reduction and compatible with the six term formation rules given at the beginning of this section. Reduction relation will be written as $v \twoheadrightarrow w$ and pronounced " v is reducible to w ", or " w is a reduct of v ". In other words, w is either v itself or may be obtained from v in finitely many *proper* steps, each of which is an immediate reduction on a subterm of a term obtained in the previous step. In section 2 we will show that this notion of reduction enjoys the Church-Rosser Confluence Property that any two reducts of the same term have a common reduct.

A polymorphic lambda term is in *normal form* if none of its subterms has an immediate reduct. w is a normal form of v if $v \twoheadrightarrow w$ and w is in normal form. The Confluence Property implies that any term can have at most one normal form. In section 3 we will prove the Strong Normalization Theorem that every sequence of proper reduction steps must be finite (Girard [72], Tait [75], Mitchell [86]) and hence that every polymorphic term has the normal form (the Normalization Property).

The reduction relation defined above is often called *beta reduction* The least equivalence relation which contains beta reduction and which is compatible with term formation is called *beta conversion* (some sources refer to it as *beta-xi conversion*). From the point of view of denotational semantics, which considers values obtained by computations, it makes sense to consider also *eta conversion*, the least equivalence relation on terms compatible with term formation that contains beta conversion, that identifies any term t of type $A \Rightarrow B$ with the term $\lambda x{:}A.(tx)$, and that identifies any term u of type $\forall p.A$ with the term $\Lambda p.(u\{p\})$.

EXAMPLE 1.5. In the example 1.1 above we discussed the polymorphic identity $\Lambda p.\lambda x{:}p.x$ of type $\forall p.p\Rightarrow p$. Term $(\Lambda p.\lambda x{:}p.x)\{A\}$ of type $A\Rightarrow A$ is immediately reducible to $\lambda x{:}A.x$ and thus for any term t of type A, $(\Lambda p.\lambda x{:}p.x)\{A\}t$ is reducible (in two steps) to t. In particular, $(\Lambda p.\lambda x{:}p.x)\{\forall p.p\Rightarrow p\}\Lambda p.\lambda x{:}p.x$ reduces to $\Lambda p.\lambda x{:}p.x$. ∎

EXAMPLE 1.6. The empty list and the list constructor in the example 1.3 are in normal form. The successor S and all polymorphic Church numerals are in normal form. Let us show that the polymorphic Church numeral 2 is the normal form of the term $S(S0)))$. The reader may reduce $S0$ to $\Lambda p.\lambda y{:}p\Rightarrow p.\lambda z{:}p.(yz)$, i.e., 1 . Now reduce $S1$ first to $\Lambda p.\lambda y{:}p\Rightarrow p.\lambda z{:}p.(y((\Lambda q.\lambda u{:}q\Rightarrow q.\lambda x{:}q.(ux))\{p\}yz))$, then to $\Lambda p.\lambda y{:}p\Rightarrow p.\lambda z{:}p.(y((\lambda u{:}p\Rightarrow p.\lambda x{:}p.(ux))yz))$, which reduces to $\Lambda p.\lambda y{:}p\Rightarrow p.\lambda z{:}p.(y((\lambda x{:}p.(yx))z))$, then to $\Lambda p.\lambda y{:}p\Rightarrow p.\lambda z{:}p.(y(yz))$. ∎

From the point of view of natural deduction, immediate reduction means that, e.g., in the case of term abstraction

$$\dfrac{\dfrac{\dfrac{\begin{array}{c}\Gamma, A\\ \vdots\end{array}}{B}}{A\Rightarrow B} \qquad \dfrac{\begin{array}{c}\Gamma\\ \vdots\end{array}}{A}}{B} \quad \text{is immediately reducible to} \quad \dfrac{\dfrac{\begin{array}{c}\Gamma\\ \vdots\end{array}}{A}\\ \vdots}{B}\ ,$$

and it also means that substitution of terms is composition of deduction trees. Reduction can thus be viewed as a simplification of deductions and hence beta normalization is of primary importance in proof theory. Its consequences include cut-elimination and the consistency of second-order arithmetic (see Girard [71,72,87b], Takeuti [87]). In fact, *higher-order* polymorphic lambda calculus F^{ω} was developed in Girard [71,72] originally as a notation for a higher-order constructive logic natural deduction system in order to prove normalization of deductions and the higher-order analogues of the consequences just mentioned. In sections 5 and 6 we shall discuss the Coquand-Huet calculus of constructions, a programming language paradigm in which higher-order deductions become programs. Together with Constable's language *Nuprl* (see Constable et al. [86]) whose theoretical basis is a somewhat different logical system studied in Martin-Löf [84], the Coquand-Huet calculus is currently a leading example in the area of computer science oriented toward programming with proofs.

2. CHURCH-ROSSER CONFLUENCE PROPERTY

THEOREM 2.1. Any two reducts of the same term have a common reduct.

PROOF: We follow the proof in Girard [72] based on the method due to Tait and to Martin-Löf in the case of untyped lambda calculus (see section 3.2 in Barendregt [84]). The key idea is to redefine reduction as the least transitive relation that includes a relation whose confluence may be established by "parallel moves".

Given $\Gamma \vdash v:A$ and $\Gamma \vdash w:A$ we define $v \blacktriangleright w$ inductively as follows, with the proviso in the rules $(\blacktriangleright 4)$ and $(\blacktriangleright 8)$ as in Type Abstraction, and with Δ a permutation of Γ in $(\blacktriangleright 6)$:

$$\frac{\Gamma \vdash t : A}{\Gamma \vdash t \blacktriangleright t : A} \qquad \blacktriangleright 1 ,$$

$$\frac{\Gamma[A/x] \vdash v \blacktriangleright w : B}{\Gamma \vdash \lambda x{:}A.t \blacktriangleright \lambda x{:}A.u : A \Rightarrow B} \qquad \blacktriangleright 2 ,$$

$$\frac{\Gamma \vdash t \blacktriangleright u : A \Rightarrow B \qquad \Gamma \vdash v \blacktriangleright w : A}{\Gamma \vdash tv \blacktriangleright uw : B} \qquad \blacktriangleright 3 ,$$

$$\frac{\Gamma \vdash v \blacktriangleright w : A}{\Gamma \vdash \Lambda p.v \blacktriangleright \Lambda p.w : \forall p.A} \qquad \blacktriangleright 4 ,$$

$$\frac{\Gamma \vdash v \blacktriangleright w : \forall p.A}{\Gamma \vdash v\{B\} \blacktriangleright w\{B\} : A[B/p]} \qquad \blacktriangleright 5 ,$$

$$\frac{\Gamma \vdash t \blacktriangleright u : A}{\Delta \vdash t \blacktriangleright u : A} \qquad \blacktriangleright 6 ,$$

$$\frac{\Gamma[A/x] \vdash t \triangleright u : B \qquad \Gamma \vdash v \triangleright w : A}{\Gamma \vdash (\lambda x{:}A.\,t)v \triangleright u[w/x] : B} \quad \triangleright 7\ ,$$

$$\frac{\Gamma \vdash v \triangleright w : A}{\Gamma \vdash (\Lambda p.\,v)\{B\} \triangleright w[B/p] : A[B/p]} \quad \triangleright 8\ .$$

Suppose that $\Gamma \vdash t \triangleright u : A$ and $\Delta \vdash t \triangleright v : A$, where Δ is a permutation of Γ. We use induction on the sum of the lenghts of derivations of the two assumptions to show that there exists a term w such that $\Gamma \vdash u \triangleright w : A$ and $\Gamma \vdash v \triangleright w : A$, and therefore also $\Delta \vdash u \triangleright w : A$ and $\Delta \vdash v \triangleright w : A$. This is trivial if one the two derivations ends with an instance of ($\triangleright$1) or ($\triangleright$6). Otherwise, first consider the case when different rules are used as the last step in the two given derivations. Then one of these two rules must be among ($\triangleright$2) - ($\triangleright$5). Without loss of generality we may assume that

$$\frac{\ldots\ \Gamma' \vdash t' \triangleright u' : A'\ \ldots}{\Gamma \vdash t \triangleright u : A}\ (\triangleright k) \qquad\qquad \frac{\ldots\ \Delta' \vdash t' \triangleright v' : A'\ \ldots}{\Delta \vdash t \triangleright v : A}\ (\triangleright n)$$

where $2 \le k \le 5$. By the induction hypothesis there is a term w' such that $\Gamma' \vdash u' \triangleright w' : A'$ and $\Gamma' \vdash v' \triangleright w' : A'$. Now use an instance of the rule ($\triangleright$n) on $\Gamma' \vdash u' \triangleright w' : A'$ instead on $\Gamma' \vdash t' \triangleright v' : A'$ and obtain $\Gamma \vdash u \triangleright w : A$. Then $\Gamma \vdash v \triangleright w : A$ may be obtained from $\Gamma' \vdash v' \triangleright w' : A'$ by using the rules ($\triangleright$2) - ($\triangleright$5). For example, let Δ and Z be permutations of Γ and consider

$$\frac{\dfrac{\begin{array}{c} Z[A/x] \vdash t \triangleright u_1 : B \\ \vdots \end{array}}{\Gamma \vdash \lambda x{:}A.\,t \triangleright \lambda x{:}A.\,u_1 : A \Rightarrow B} \qquad \Gamma \vdash v \triangleright w_1 : A}{\Gamma \vdash (\lambda x{:}A.\,t)v \triangleright (\lambda x{:}A.\,u_1)w_1 : B} \quad (\triangleright 3)$$

and

$$\frac{\Delta[A/x] \vdash t \triangleright u_2 : B \qquad \Delta \vdash v \triangleright w_2 : A}{\Delta \vdash (\lambda x{:}A.\,t)v \triangleright u_2[w_2/x] : B} \quad (\triangleright 7).$$

Then, first of all, $\Gamma[A/x] \vdash t \triangleright u_1 : B$ by ($\triangleright$6) and we may still apply the induction hypothesis. Therefore there exist polymorphic terms u_3 and w_3 such that

$$\Gamma[A/x] \vdash u_1 \triangleright u_3 : B, \qquad \Gamma[A/x] \vdash u_2 \triangleright u_3 : B,$$

$$\Gamma \vdash w_1 \triangleright w_3 : A, \quad \text{and} \quad \Gamma \vdash w_2 \triangleright w_3 : A.$$

By ($\triangleright$7) the first and the third of these conclusions yield

$$\frac{\Gamma[A/x] \vdash u_1 \triangleright u_3 : B \qquad \Gamma \vdash w_1 \triangleright w_3 : A}{\Gamma \vdash (\lambda x{:}A.\, u_1)w_1 \triangleright u_3[w_3/x] : B},$$

while the second and the fourth yield $\Gamma \vdash u_2[w_2/x] \triangleright u_3[w_3/x] : B$ by ($\triangleright$2) - ($\triangleright$6). Furthermore, by ($\triangleright$6) we may obtain the analogues with Δ instead of Γ.

In the case when the last steps in the given two derivations are instances of the same rule, again use ($\triangleright$2) - ($\triangleright$6) and the induction hypothesis.

We have shown that the relation $\triangleright$ is confluent. A simple diagram chase indicated by

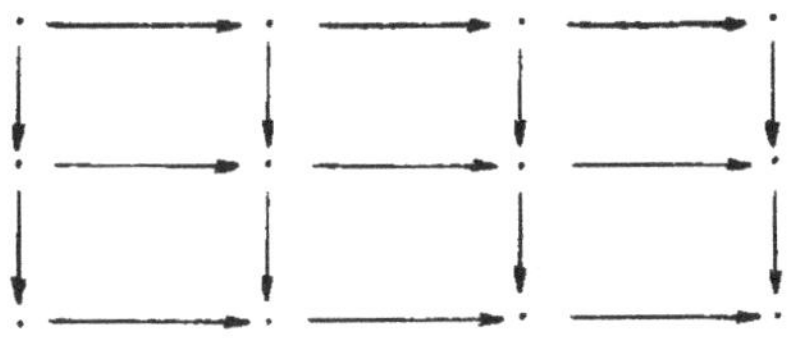

now shows that beta reduction is confluent, because u is reducible to v (both of type A) iff there is a context Γ and polymorphic terms $t_1, \ldots, t_n$ so that $\Gamma \vdash u \triangleright t_1 : A, \ldots, \Gamma \vdash t_n \triangleright v : A$. ∎

A general theory of reduction in term rewriting systems has been developed in Lévy [78], Huet and Lévy [79], Huet [80], and Klop [80].

3. *STRONG NORMALIZATION*

The Strong Normalization Theorem, that every sequence of proper reductions terminates in finitely many steps, was originally proved in Girard [71,72]. As a consequence, every term has the normal form

(Normalization Property), which must be unique by the Confluence Property. In computer science normalization is important as a theoretical basis for program verification mechanisms such as type checking. For example, current proofs of decidability of type-checking in sophisticated type disciplines such as the Coquand-Huet calculus of constructions rely on normalization, see section 5 below and Huet [87]. Normalization is also used in various conservative extension results about integrating data type algebras into polymorphic lambda calculus, e.g. Breazu-Tannen and Meyer [87], Breazu-Tannen [88]. Besides, the techniques used in proving normalization are related to the techniques used in establishing some desirable properties concerning the relationships between operational and denotational semantics, see Meyer [88], Moggi [88], and Plotkin [85].

We present a version of Girard's proof of strong normalization given in Mitchell [86] (see also Tait [75]). It deals with *untyped lambda terms*, which, as the reader will recall, are defined inductively as follows:

a) Assume a countably infinite collection of variables, each of which is an untyped lambda term (term, in short),
b) If a is a term and x is a variable, then $\lambda x.a$ is a term,
c) If a and b are terms, then ab is a term.

Free occurrences of a variable in a term, substitution of terms, and reduction are defined in the same manner as above (but without restrictions due to types). In particular, we identify terms up to renaming of bound variables (alpha conversion). Reduction is often called *beta reduction*. Beta reduction enjoys the confluence property and will again be denoted as $a \twoheadrightarrow b$. Normalization fails for untyped lambda terms. For example, $(\lambda x.xx)(\lambda x.xx)$ is its only reduct but it is not in normal form.

(The reader is referred to Barendregt [84] and Hindley & Seldin [86] for a detailed study of the untyped lambda calculus. Further references are Huet [87], Cousineau et al. [86] , Curien [86], and Lambek and Scott [86].)

We write t , u , v , ... for polymorphic typed lambda terms and a , b , c , ... for untyped lambda terms. We say that an untyped lambda term c is *strongly normalizable* if every sequence of proper reductions of c must terminate in finitely many steps (in the normal form of c).

A *saturated set* is a set S of strongly normalizable untyped lambda terms such that:

a) $xa_1 \dots a_n$ is in S for any variable x and any strongly normalizable $a_1, \dots, a_n$ (applications are associated to the left, as usual),

b) if b_0 is strongly normalizable and $(a[b_0/x])b_1 \dots b_n$ is in S, then $(\lambda x.\, a)b_0 b_1 \dots b_n$ is in S.

(The reader may check that it suffices to require these properties only for strongly normalizable a, $b_1, \dots, b_n$.)

Polymorphic types are interpreted as saturated sets:

$\|p\|$ is an arbitrary saturated set for any type variable p,
$\|A \Rightarrow B\| = \{c:\ c$ strongly normalizable and for any $a \in \|A\|$, $ca \in \|B\|\}$,
$\|\forall p.\, A\| = \bigcap \{\|A\|:\ \|p\|$ any saturated set$\}$.

We must verify that these sets are indeed saturated. Condition b) is not quite obvious for the set defined by the second clause. Suppose b_0 is strongly normalizable and $(d[b_0/x])b_1 \dots b_n \in \|A \Rightarrow B\|$, the latter meaning that $(d[b_0/x])b_1 \dots b_n$ is strongly normalizable and $(d[b_0/x])b_1 \dots b_n a \in \|B\|$ for $a \in \|A\|$. Then $(\lambda x.\, d)b_0 b_1 \dots b_n$ is strongly normalizable and because $\|B\|$ is a saturated set, $(\lambda x.\, d)b_0 b_1 \dots b_n a \in \|B\|$. Thus $(\lambda x.\, d)b_0 b_1 \dots b_n \in \|A \Rightarrow B\|$.

Any context Γ is interpreted as the (finite) list $\|\Gamma\|$ of saturated sets that interpret the types occurring in Γ.

Polymorphic terms are interpreted by erasing types. More precisely:

$(\lambda x{:}A.\, t)^- = \lambda x.\, t^-$,
$(tu)^- = t^- u^-$,
$(\Lambda p.\, t)^- = t^-$,
$(t\{A\})^- = t^-$.

We now show that this interpretation is sound. If $\vec{x}$ is a list of variables $x_1 \ldots x_n$ and $\vec{a}$ is a list of strongly normalizable untyped lambda terms, then we write $t^-[\vec{a}/\vec{x}]$ for the result of simultaneous substitution of a_k for free x_k in t^-, $1 \leq k \leq n$. (Nothing is done with a_k if x_k is not free in t^-.) We remind the reader that the empty conjunction is true.

LEMMA 3.1. Let $\Gamma \vdash t:A$ and $\vec{a} \in \|\Gamma\|$. Then $t^-[\vec{a}/\vec{x}] \in \|A\|$.

PROOF: By induction on the length of derivation of typing judgments (see the inductive definition of derivable typing judgments in section 1). The only problematic case is Lambda Abstraction:

$$\frac{\Gamma[A/x] \quad \vdash \quad t:B}{\Gamma \quad \vdash \quad \lambda x{:}A.\,t \;:\; A \Rightarrow B} \quad .$$

We suppress parameters other than x. By the induction hypothesis, for any $a \in \|A\|$, $t^-[a/x] \in \|B\|$ and thus the untyped lambda term $t^-[a/x]$ is strongly normalizable. Because $\|A\|$ is a type set it must contain the variable x itself. Hence $t^- \in \|B\|$ and thus t^- must be strongly normalizable. Therefore $\lambda x.t^-$ is strongly normalizable. Of course, $\lambda x.t^-$ is $(\lambda x{:}A.\,t)^-$. It remains to be shown that $(\lambda x{:}A.\,t)^-a$ belongs to the type set $\|B\|$ for any $a \in \|A\|$. But $\|B\|$ is a saturated set, $t^-[a/x] \in \|B\|$ by the induction hypothesis, and a is strongly normalizable because $a \in \|A\|$ and saturated sets consist only of strongly normalizable terms. Now property b) of the saturated set $\|B\|$ yields $(\lambda x{:}A.\,t)^-a \in \|B\|$. ∎

COROLLARY 3.1. The type erasure of every second-order polymorphic lambda term is strongly normalizable in the untyped lambda calculus. ∎

We finish the proof of strong normalization for second-order polymorphic lambda calculus by noting the following two immediate facts:

LEMMA 3.2. If t is immediately reducible to u in polymorphic lambda calculus by a reduction step on a type application $(\Lambda p.\,w)\{A\}$ then the number of occurrences of the symbol Λ in u is one less than in t. ∎

LEMMA 3.3. If t is reducible to u in polymorphic lambda calculus, then t^- is reducible to u^- in untyped lambda calculus. If t is reducible to u in polymorphic lambda calculus by a finite sequence that includes an immediate reduction on a term application $(\lambda x{:}A.\,w)v$, then t^- is reducible to u^- in untyped lambda calculus by a finite sequence that includes at least one step $(\lambda x.\,a)c \twoheadrightarrow a[c/x]$. ∎

Therefore:

THEOREM 3.1. Every sequence of proper reductions of a second-order polymorphic lambda term must terminate in finitely many steps.

PROOF: Corollary 3.1 and Lemmas 3.2 and 3.3 ∎

A somewhat simpler version of this proof that yields only normalization may be found in Scedrov [87b].

It has been recently shown in Giannini et al. [87,88] that the converse of Corollary 3.1 does not hold. They give an example of a strongly normalizable untyped lambda term that is not the type erasure of any second-order polymorphic lambda term.

One of the consequences of normalization is the exact description of the class of recursive functions representable in second-order polymorphic lambda calculus. It is an old result of Church and Kleene that a (total) function from natural numbers to natural numbers is representable in the untyped lambda calculus iff it is recursive (see e.g. Barendregt [84]). Girard [72] showed that the representability in second-order polymorphic lambda calculus requires in addition that a recursive function must be provably total in second-order arithmetic.

More precisely, let $\overline{n}$ be the polymorphic numeral representing natural number n. A function $f\colon N \longrightarrow N$ is said to be representable in second-order polymorphic lambda calculus if there exists a polymorphic term t of type $Nat \Rightarrow Nat$, without free variables, such that for every natural number n, the term $t\overline{n}$ is reducible to the polymorphic numeral $\overline{f(n)}$. Girard [72] showed that $f\colon N \longrightarrow N$ is representable in second-order polymorphic lambda calculus iff it is recursive and provably total in second-order arithmetic.

On the other hand, Krivine [87] gives an example of an *algorithm* for computing the minimum of two natural numbers (recall that the function described is primitive recursive), which is given by an untyped lambda term that does not arise as the type erasure of a polymorphic term of type $Nat \Rightarrow Nat$. Thus, while second-order polymorphic lambda calculus suffices to represent *an* algorithm for any recursive function that is provably total in second order arithmetic (Girard's result), this calculus is already not sufficient to represent *any* algorithm for rather innocuous recursive functions.

4. AN OVERVIEW OF TOPICS IN SEMANTICS

Mathematical semantics is an important component of programming language design. One of the roles of mathematical semantics is in providing a conceptual guide to a consistent formulation of syntax and thus help avoiding ad hoc syntactic extensions and enrichments. On the other hand, a good semantics would also provide a clear framework which would make it possible to integrate polymorphism with other desirable features of programming languages such as coercion, inheritance, dependent types, recursion, and concurrency.

A first, primitive attempt is to consider types as *sets* and typing judgments $\Gamma \vdash t{:}A$ as *functions* from the cartesian product of sets that interpret types in Γ to the set that interprets A. There is an obvious problem with interpreting type abstraction, which turns out to be insurmountable. It is shown in Reynolds [84] (see also Reynolds and Plotkin [88]) that there are no nontrivial ordinary set-theoretic interpretations of second-order polymorphic lambda calculus.

The situation is radically different as soon as one allows *constructive* logic as a logical framework for set theory. Not only are then plenty of interesting set-theoretic interpretations, but Pitts [87] has also shown the completeness theorem for such models. General notion of a model of second-order polymorphic lambda calculus (with eta conversion) is studied in Bruce et al. [8x], Seely [87], and Meseguer [88].

One important line of research in models for polymorphism stems from the work in domain theory in Scott [72, 76, 82], Smyth and Plotkin [82] in which data types are modelled as certain partially ordered sets with a least element and with directed suprema, while programs are modelled as monotone functions that preserve directed suprema, often called continuous functions (see Gunter and Scott [88]). The first domain-theoretic model for polymorphic lambda calculus was the $P\omega$ model given in McCracken [79]. The finitary projection model was studied in Amadio et al. [86].

A new impetus to domain-theoretic study of models for polymorphism is given by Girard [86], where types are treated as (n-ary) functors on the category of certain *coherent* domains and embedding-projection pairs and terms t of type F are certain uniform families of elements t_X of domains F(X) indexed by *all* (n-tuples of) domains X. It is shown that every such family is uniquely determined by the subfamily indexed by a certain *set* of finite coherent domains, and that such subfamilies are terms of the type abstraction $\forall p.F$. Girard's idea was further investigated in the contexts of dI-domains and of Scott domains in Coquand et al. [88a, b], motivated in part by considerations of a notion of a direct sum of domains. A curious aspect of both the coherent semantics and the dI-domains semantics is the connection with semantics of concurrency. On the one hand, the work on dI-domains for polymorphism arose in part from Winskel's study of *event structures* as a semantics for concurrency. On the other hand, coherent domains were intended as a setting for polymorphism, but it turned out that a more fundamental structure of *linear logic* was being modelled (Girard [87a]) and deductions in linear logic resemble concurrent computations in much the same way that deductions in constructive logic resemble sequential computations (propositions as types paradigm).

Another semantic context for polymorphism that has recently received a lot of attention is the Realizability Universe (also called the Effective Topos), see Carboni et al. [88], Longo and Moggi [87], Rosolini [86], Scott [87], Hyland et al. [87], Freyd and Scedrov [87]. It as a model for constructive set theory and higher-order logic with a special feature, originally emphasized by Moggi, that this model contains a nontrivial set $\mathbf{M}$ closed under finite cartesian products, under formation of sets of functions and under products of those families of elements of **M** that are defined intrinsically in the Realizability Universe. In addition, if a set belongs to **M** then all of its subsets in the Realizability Universe must belong to **M**.

The elements of **M** are called *modest sets*. Viewed from the inside the Realizability Universe, modest sets are exactly the quotients of subsets of the natural numbers by double negation stable equivalence relations. In the Realizability Universe, modest sets and arbitrary functions between them provide a setting for a set-theoretic interpretation of polymorphic lambda calculus in which a function type $A \Rightarrow B$ is interpreted as a set of all functions from a modest set A to a modest set B ($A \Rightarrow B$ is again modest) and type abstraction $\forall p. A$ is interpreted as the product of a family of modest sets $A[B/p]$ indexed by *all* modest sets B (this product is again modest).

Viewed from the outside, modest sets are symmetric, transitive binary relations on the natural numbers, often called partial equivalence relations and abbreviated as *pers*. The semantics of polymorphism just described is an example of PER semantics (see Breazu-Tannen and Coquand [87]). A per map from R to S is named by a partial recursive function f such that for any n and k, if $n\ R\ k$ then $f(n)$ and $f(k)$ must be defined and $f(n)\ S\ f(k)$. Two such partial recursive functions f and f' are understood as naming the same per map iff for any n and k, if $n\ R\ k$ then $f(n)\ S\ f'(k)$. Cartesian product of pers $R \cdot R'$ may be described by means of a primitive recursive coding of pairs of natural numbers: $\langle n,k\rangle\ R \cdot R'\ \langle i,j\rangle$ iff $n\ R\ i$ and $k\ R'\ j$. Per $R \Rightarrow S$ may be described by: $e\ R \Rightarrow S\ e'$ iff e and e' are numerical codes of partial recursive functions that name the same per map. Per $\forall p. R$ is simply the intersections of pers $R[P/p]$ for all pers P. The amazing fact that this intersection is a product in the Realizability Universe is a consequence of an equally amazing fact that in the Realizability Universe, every function (indeed, every total relation) from the set of modest sets **M** to the set of natural numbers must be constant. In addition, it has been shown by Hyland and by Freyd that the type of polymorphic natural numbers $\mathit{Nat} = \forall p. ((p \Rightarrow p) \Rightarrow p \Rightarrow p)$ is interpreted exactly as the per given by the set of natural numbers with the ordinary equality.

A typing judgment $\Gamma \vdash t:A$ is interpreted as a per map from the cartesian product of pers that interpret types in Γ to the per that interprets A. In interpreting lambda abstraction, use the s-m-n theorem from recursion theory. In interpreting application use the Kleene bracket application of a numerical code of a partial recursive function to a numerical argument. This interpretation of polymorphic lambda calculus is in fact Girard's HEO_2 interpretation given in Girard [72], but the key departure in the recent research mentioned above is that in the Realizability Universe, polymorphism is integrated within standard mathematical operations on sets. This feature of the Realizability Universe raises the possibility of combining the Scott-Strachey fixed-point theory of computation (Scott [76, 82]) with the propositions-as-types computational paradigm in a single framework (Scott [87]). We also note recent work of Bruce and Longo which gives a semantics of type coercion in the Realizability Universe (Bruce and Longo [88]) by means of inclusions of pers.

One current topic of investigation is a semantic formulation of the notion of parametricity mentioned in the introduction, which is the very notion Reynolds intended to capture in the syntax of second-order polymorphic lambda calculus. Reynolds [83] points out that it is not enough to simply interpret typing judgments, but that the interpretation of an abstraction $\forall p.A$ must consist only of the parametric functions, those that are invariant under the relations between values of p. This implies that in a semantics of parametric polymorphism $\forall p.A$ should not be interpreted as a product (i.e. as consisting of all functions). Mitchell and Meyer [85] treat relations between type values as second-order logical relations and prove that in the term model of second-order polymorphic lambda calculus all elements of values of universal types $\forall p.A$ are invariant under all second-order logical relations (since in the term model all elements are definable by terms).

Reynolds' requirement may also be addressed to some extent by defining parametricity semantically as certain equational conditions that express uniformity under all functional substitutions, and then interpreting $\forall p.A$ as consisting only of the uniform families. A refinement of PER semantics is obtained in this way in Bainbridge et al. [87]. Freyd et al. [88] use this approach to obtain a refinement of the coherent semantics (Girard [86]) in the category of coherent spaces and *linear maps*, one of whose consequences is a notion of a sum of coherent spaces. In general, this approach yields a systematic way of adding new desirable equations between terms that do not follow from eta conversion.

5. CALCULUS OF CONSTRUCTIONS

Calculus of constructions is an extension of the Girard-Reynolds polymorphic lambda calculus that combines higher-order polymorphic types with dependent products. In calculus of constructions the expressions for *proofs* and *assertions* (or synonymously: for *programs* and *specifications*) may be given uniformly in all higher orders and types. Several versions of this kind of lambda formalism have been studied and implemented by T. Coquand and G. Huet (see Huet [87], Coquand & Huet [85, 87], Coquand [85], and Mohring [86]). Our presentation mostly follows the presentation in Huet [87] with one difference in terminology: our **Kind** is Huet's **Type**. The terms are defined inductively:

$$\mathbf{Prop} \mid \mathbf{Kind} \mid x \mid [x:A]B \mid (x:A)B \mid (AB) .$$

Abstraction over A is denoted by square brackets $[x:A]$. Product and quantification over A are denoted by parentheses $(x:A)$. We often write κ for either **Prop** or **Kind**. As in Huet [87], we here consider variables as notations for de Bruijn indices (see de Bruijn [72]) and thus we do not need alpha conversion.

Contexts are finite lists of bindings of variables: either empty or the expressions of the form $[x_1:A_1] \ldots [x_n:A_n]$, where x_k is a variable and A_k is a term for each $1 \leq k \leq n$. If Γ is such a nonempty context, we write Γ_k for the term A_k .

Judgments are expressions of the form:

$$\Gamma \vdash A : B ,$$

or:

$$\Gamma \vdash A = B ,$$

where Γ is a context and A and B are terms.

Derived judgments and *valid contexts* are defined simultaneously by the following rules of inference, in all of which Γ is assumed to be a valid context. In addition, in the second rule, variable x may not occur in Γ ; in the third rule, Γ must be nonempty.

The empty context is valid.

$$\frac{\Gamma \vdash A:\kappa}{\Gamma[x:A] \quad \text{valid}}$$

Var $$\Gamma \vdash x_K : \Gamma_K$$

Prop $$\Gamma \vdash \mathbf{Prop} : \mathbf{Kind}$$

Prod $$\frac{\Gamma \vdash A:\kappa \qquad \Gamma[x:A] \vdash B : \mathbf{Kind}}{\Gamma \vdash (x:A)B : \mathbf{Kind}}$$

Quant $$\frac{\Gamma \vdash A:\kappa \qquad \Gamma[x:A] \vdash B : \mathbf{Prop}}{\Gamma \vdash (x:A)B : \mathbf{Prop}}$$

Abstr $$\frac{\Gamma \vdash A:\kappa \qquad \Gamma[x:A] \vdash B:\kappa' \qquad \Gamma[x:A] \vdash M:B}{\Gamma \vdash [x:A]M : (x:A)B}$$

Appl $$\frac{\Gamma \vdash (x:A)B : \kappa \qquad \Gamma \vdash L : (x:A)B \qquad \Gamma \vdash N:A}{\Gamma \vdash (LN) : B\{N/x\}}$$

Equal $$\frac{\Gamma \vdash M:A \qquad \Gamma \vdash B:\kappa \qquad \Gamma \vdash A = B}{\Gamma \vdash M:B}$$

Refl $$\frac{\Gamma \vdash M:A}{\Gamma \vdash M = M}$$

Sym $$\frac{\Gamma \vdash M = N}{\Gamma \vdash N = M}$$

Trans $$\frac{\Gamma \vdash L = M \qquad \Gamma \vdash M = N}{\Gamma \vdash L = N}$$

AbsEq
$$\frac{\Gamma \vdash A = A' \qquad \Gamma[x{:}A] \vdash M = M'}{\Gamma \vdash [x{:}A]M = [x{:}A']M'}$$

QuantEq
$$\frac{\Gamma \vdash A = A' \qquad \Gamma[x{:}A] \vdash B = B'}{\Gamma \vdash (x{:}A)B = (x{:}A')B'}$$

ApplEq
$$\frac{\Gamma \vdash (LN) : A \qquad \Gamma \vdash L = L' \qquad \Gamma \vdash N = N'}{\Gamma \vdash (LN) = (L'N')}$$

Beta
$$\frac{\Gamma[x{:}A] \vdash M{:}B \qquad \Gamma \vdash N{:}A}{\Gamma \vdash ([x{:}A]M\ N) = M\{N/x\}}$$

Constructions consist of finitely many successive applications of the inference rules. From now on we assume that all exhibited judgments are derivable and that Γ denotes a valid context.

The following three lemmas may be shown by induction on derivation of judgments:

LEMMA 5.1. If $\Gamma \vdash M{:}A$, then either $\Gamma \vdash A = \mathbf{Kind}$, or $\Gamma \vdash A{:}\mathbf{Kind}$ or $\Gamma \vdash A{:}\mathbf{Prop}$. ∎

Referring to the three cases given by the lemma, we say that M is either a *Γ-kind* or a *Γ-proposition* or a *Γ-proof* . We often omit the prefix Γ when discussing the cases in a single construction. Propositions will often also be called types. *Warning: in Huet [87] and Scedrov [87a] types refer to what we here call kinds.* We introduce this change in terminology in order to be consistent with the terminology in the literature on second-order polymorphic lambda calculus, which is maintained in Cardelli's experimental programming language *Quest* (Cardelli [88]).

LEMMA 5.2. If $\Gamma \vdash M : A$ and $\Gamma \vdash M = N$, then $\Gamma \vdash N : A$. ∎

Moreover, in the presence of eta conversion:

LEMMA 5.3. If $\Gamma \vdash M : A$ and $\Gamma \vdash M : B$, then $\Gamma \vdash A = B$. ∎

We use the following abbreviations for $(x{:}A)B$ if x does not occur in B : when both A and B are propositions we write $A \Rightarrow B$, else $A \dashrightarrow B$. In this notation we associate to the right.

Beta reduction in calculus of constructions is defined similarly to the conversion rules indicated by the equality judgments, so that the relationship between beta reduction and beta conversion is the same as in second-order polymorphic lambda calculus. A proof of strong normalization for calculus of constructions indicated in Huet [87] is analogous to the version of Girard's proof of strong normalization for second-order polymorphic lambda calculus given in Tait [75] and Mitchell [86] and described in section 3 above.

THEOREM 5.1. Every sequence of beta reductions in calculus of constructions terminates in finitely many steps. ∎

An important consequence of normalization is *decidability of type checking*:

THEOREM 5.2. Given Γ and M , it is decidable whether or not there exists a term A such that $\Gamma \vdash M : A$. Moreover, if the answer is positive, we can compute effectively such an A . ∎

Note that decidability of type-checking is trivial for polymorphic lambda calculus because term abstractions are typed and because types cannot depend on terms. The latter is not the case in calculus of constructions.

6. AN INTERPRETATION OF CALCULUS OF CONSTRUCTIONS

We discuss the interpretation of calculus of constructions given in Scedrov [87a] where typing of terms is treated as set membership, i.e., typings "term M has type A in context Γ" are interpreted by assigning a set $ext(\Gamma)$ to Γ and certain maps $\|M\|$ and $\|A\|$ to M and A so that for each c in $ext(\Gamma)$, $\|M\|(c)$ belongs to $\|A\|(c)$. On the one hand, this interpretation is related to the HEO_n semantics for finite higher-order lambda calculi (see Girard [72, ch. II.5.]). On the other hand, our interpretation may be viewed as an interpretation of calculus of constructions in the Realizability Universe, which extends PER semantics of second-order polymorphic lambda calculus mentioned in section 4 above. This recursive realizability interpretation of calculus of constructions has also been outlined by M. Hyland and by D. Scott. Other interpretations, based on category theory, are given in Hyland and Pitts [87] and in Lamarche [87].

Given any derived judgment $\Gamma \vdash M:A$, we specify a set $ext(\Gamma)$ (the *extent* of Γ) and a mapping $\|M\|$ defined on $ext(\Gamma)$ (the value of M). This will be done in such a way that whenever $\Gamma \vdash M:A$ and $\Gamma \vdash A:\kappa$ (recall the three cases in Lemma 5.1) it will be the case that for any $c \in ext(\Gamma)$, $\|M\|(c) \in \|A\|(c)$. Moreover, derivable equality of terms will be interpreted as ordinary equality of their values given by their types.

Given a valid context Γ of the form $[x_1:A_1]\ldots[x_n:A_n]$, $n \geq 0$, we count the number $\#\Gamma$ of non-kinds among $A_1, \ldots, A_n$ and define a canonical injection $j_\Gamma: \{1,\ldots,\#\Gamma\} \longrightarrow \{1,\ldots,n\}$ as follows: $\#\_ = 0$, $j\_ = 0$; if Γ' is $[x_1:A_1]\ldots[x_{n-1}:A_{n-1}]$ and A_n is a type, then $\#\Gamma = \#\Gamma'$ and $j_\Gamma(K) = j_{\Gamma'}(K)+1$, else $\#\Gamma = (\#\Gamma')+1$, $j_\Gamma(1) = 1$, and $j_\Gamma(K+1) = j_{\Gamma'}(K)+1$. We will often confuse Γ with a derived judgment $\Gamma \vdash \mathbf{Prop} : \mathbf{Kind}$ and thus with a derived judgment $\vdash (x_1:A_1)\ldots(x_n:A_n)\mathbf{Prop} : \mathbf{Kind}$. Every kind is derivably equal to one of this form.

Let $\|\mathbf{Prop}\|$ be the collection of all quotients of equivalence relations on sets of natural numbers. We proceed by induction on the definition of constructions and valid contexts. Simultaneously, we show that a value of a Γ-proof M is uniquely determined by a partial recursive function φ_M of $\#\Gamma$ arguments that preserves the relevant symmetric transitive relations. We say that φ_M *realizes* M.

A few words about notation: for A/R in $\|\mathbf{Prop}\|$, let $dom(A/R) = A$ $rel(A/R) = R$. Given Q in $\|\mathbf{Prop}\|$ and a map $F: Q \dashrightarrow \|\mathbf{Prop}\|$, we write $\tilde{n}$ for the equivalence class of n in Q. Let $\Pi rec_{a \in Q} F(a)$ be the quotient in $\|\mathbf{Prop}\|$ given as follows: first consider the set of numerical codes of partial recursive functions f such that if $n\ rel(Q)\ m$, then $f(n)$ and $f(m)$ are defined, and $f(n)$ and $f(m)$ are related in $F(\tilde{n})$. Note that if $n\ rel(Q)\ m$, $F(\tilde{n}) = F(\tilde{m})$. We consider two such codes e, e' equivalent if $\{e\}(n)$ and $\{e'\}(m)$ are related in $F(\tilde{n})$ whenever $n\ rel(Q)\ m$.

In the following definition, c is assumed to be in $ext(\Gamma)$.

Rules for valid contexts: Let the extent of a blank context be a one element set. Let $\mathrm{ext}(\Gamma[x:A])$ consist of all ordered pairs $\langle c,a\rangle$ such that $c \in \mathrm{ext}(\Gamma)$ and $a \in \|A\|(c)$.

Var: Since $\mathrm{ext}(\Gamma)$ is the set of n-tuples $\langle a_1,\ldots,a_n\rangle$ such that $a_{m+1} \in \|A_m\|(a_1,\ldots,a_m)$ for $1 \leq m \leq n$, let $\|x_k\|(a_1,\ldots,a_n) = a_k$. If Γ_k is a proposition in the context $[x_1:A_1] \ldots [x_{k-1}:A_{k-1}]$, then let the required partial recursive function of several arguments be the projection of $\#\Gamma$ arguments to that coordinate which is the inverse image of k under j_Γ .

Prod: Let $\|(x:A)B\|(c) = \Pi_{a\in\|A\|(c)} \|B\|(c,a)$, an ordinary product of sets. Recall that a product over the empty set has exactly one element.

Quant: *Case 1:* κ is **Kind** . Let $\mathrm{dom}(\|(x:A)B\|(c))$ be the intersection of $\mathrm{dom}(\|B\|(c,a))$ for all $a \in \|A\|(c)$. Let $\mathrm{rel}(\|(x:A)B\|(c))$ be the intersection of $\mathrm{rel}(\|B\|(c,a))$ for all $a \in \|A\|(c)$.
Case 2: κ is **Prop** . Let $\|(x:A)B\|(c) = \Pi rec_{a\in\|A\|(c)} \|B\|(c,a)$.

Abstr: *Case 1:* κ' is **Kind** . Let $\|[x:A]M\|(c)$ be the assignment defined by $\|[x:A]M\|(c)(a) = \|M\|(c,a)$ for all $a \in \|A\|(c)$.
Case 2: κ' is **Prop**, κ is **Kind**. Then $\#\Gamma[x:A] = \#\Gamma$. We know that $\|M\|(c,a)$ = the equivalence class of $\varphi_M(n_1,\ldots,n_{\#\Gamma})$. Now simply let $\varphi_{[x:A]M} = \varphi_M$. *Case 3:* κ' and κ are both **Prop** . Then $\#\Gamma[x:A] = \#\Gamma + 1$. We define $\varphi_{[x:A]M}$ by the s-m-n theorem from recursion theory. For each fixed $n_1, \ldots, n_{\#\Gamma}$, let $\varphi_{[x:A]M}(n_1,\ldots,n_{\#\Gamma})$ be the numerical code (given by the s-m-n theorem) of the unary partial recursive function that maps k to $\varphi_M(n_1,\ldots,n_{\#\Gamma},k)$. $\varphi_{[x:A]M}$ preserves the required relations because, according to *Quant*, *Case 2*, $\|(x:A)B\|$ is of the form Πrec .

Appl: *Case 1:* κ is **Kind**. Let $\|LN\|$ be given by evaluation, that is: $\|LN\|(c) = \|L\|(c)(\|N\|(c))$. *Case 2:* κ is **Prop** and $\Gamma \vdash A$:**Kind**. Let $\varphi_{LN} = \varphi_L$. *Case 3:* κ is **Prop** and $\Gamma \vdash A$:**Prop** . Then let $\varphi_{LN}(n_1,\ldots,n_{\#\Gamma})$ be obtained by computing the value of the partial recursive function of numerical code $\varphi_L(n_1,\ldots,n_{\#\Gamma})$ on input $\varphi_N(n_1,\ldots,n_{\#\Gamma})$ (when all data are defined) . In recursion-theoretic notation, $\varphi_{LN}(n_1,\ldots,n_{\#\Gamma}) \simeq \{\varphi_L(n_1,\ldots,n_{\#\Gamma})\}(\varphi_N(n_1,\ldots,n_{\#\Gamma}))$. ∎

The soundness of this interpretation is verified by induction on constructions:

THEOREM 6.1. Let $\Gamma \vdash M{:}A$ and $\Gamma \vdash A{:}\kappa$. Then for any any $c \in ext(\Gamma)$, $\|M\|(c) \in \|A\|(c)$. Furthermore, if $\Gamma \vdash A = B$ then for any $c \in ext(\Gamma)$, $\|A\|(c) = \|B\|(c)$ and if $\Gamma \vdash M = N$ then for any $c \in ext(G)$, $\|M\|(c) = \|N\|(c)$ in $\|A\|$. ∎

The consistency of the calculus follows because $\|(x{:}\mathbf{Prop})x\|$ is empty:

COROLLARY 6.1. There is no term M such that $\vdash M : (x{:}\mathbf{Prop})x$. ∎

We have mentioned in section 4 that it has been shown by Hyland and by Freyd that in the recursive realizability (PER) semantics the second-order polymorphic type of natural numbers $\forall p.((p \Rightarrow p) \Rightarrow p \Rightarrow p)$ is interpreted exactly as the set of natural numbers with the ordinary equality thought of as a per. This result directly translates into our setting (recall the notation introduced after Lemma 5.3):

PROPOSITION 6.1. $\|(x{:}\mathbf{Prop})((x \Rightarrow x) \Rightarrow x \Rightarrow x)\|$ is the set of natural numbers with the ordinary equality. ∎

This result may be thought of as a stronger version of consistency.

EXAMPLE 6.1. Consider the following construction of the polymorphic identity:

1.		$\vdash$	$\mathbf{Prop} : \mathbf{Kind}$,
2.	$[x{:}\mathbf{Prop}]$		valid ,
3.	$[x{:}\mathbf{Prop}]$	$\vdash$	$x{:}\mathbf{Prop}$,
4.	$[x{:}\mathbf{Prop}]\ [y{:}x]$		valid ,
5.	$[x{:}\mathbf{Prop}]\ [y{:}x]$	$\vdash$	$y{:}x$,
6.	$[x{:}\mathbf{Prop}]\ [y{:}x]$	$\vdash$	$x{:}\mathbf{Prop}$,
7.	$[x{:}\mathbf{Prop}]$	$\vdash$	$[y{:}x]y : (y{:}x)x$ (*Abstr* on 6, 6, 5) ,
8.	$[x{:}\mathbf{Prop}]$	$\vdash$	$(y{:}x)x : \mathbf{Prop}$ (*Quant* on 3, 6) ,
9.		$\vdash$	$(x{:}\mathbf{Prop})(y{:}x)x : \mathbf{Prop}$ (*Quant* on 1, 8) ,
10.		$\vdash$	$[x{:}\mathbf{Prop}][y{:}x]y : (x{:}\mathbf{Prop})(y{:}x)x$ (*Abstr*) .

We may interpret this construction "as it unfolds". Recall that $\|\mathbf{Prop}\|$ is the set of all quotients of equivalence relations on subsets of natural numbers. Also recall that the extent of the empty context is a one-element set. We proceed along the construction:

2. ext([x:**Prop**]) = $\|\mathbf{Prop}\|$,
3. $\|x\|$ is the identity function on $\|\mathbf{Prop}\|$,
4. ext([x:**Prop**] [y:x]) consists of all ordered pairs $\langle Q, a\rangle$ where $Q \in \|\mathbf{Prop}\|$ and $a \in Q$,
5. # [x:**Prop**] [y:x] = 1 , $\|y\|(Q, a) = a$, φ_y is the identity function on the natural numbers,
6. $\|x\|(Q, a) = \|x\|(Q) = Q$,
7. # [x:**Prop**] = 0 , $\|[y:x]y\|(Q)$ = the s-m-n numerical code of the identity function,
8. $\|(y:x)x\|(Q) = \Pi rec_{a \in Q}\, Q$, that is: the set I_Q of numerical codes of partial recursive functions that map dom(Q) to itself and preserve rel(Q) ; modulo the equivalence relation R_Q on I_Q : $e\ R_Q\ e'$ iff n rel(Q) m implies $\{e\}(n)$ rel(Q) $\{e'\}(m)$,
9. $\|(x:\mathbf{Prop})(y:x)x\|$ is the quotient I/R , where I is the intersection of all I_Q 's and R is the intersection of all R_Q 's ,
10. $\|[x:\mathbf{Prop}][y:x]y\|$ is the code of the identity used in 7. ∎

EXAMPLE 6.2. We discuss the recursive realizability interpretation of example 5.2.3 in Huet [88]. The example is a good illustration of the importance of equality judgments. Given a kind A in a valid context Γ', we will construct a proof of the proposition that the intersection of a class of predicates on A is included in any predicate of the class. Let us begin by defining the inclusion of predicates on A . Recall the notation introduced after Lemma 5.3. Let *Subset* be the term:

[P: A --> **Prop**] [Q: A --> **Prop**] (x:A) (Px)⇒(Qx) .

We first show that:

Γ' ⊢ *Subset* : (A --> **Prop**) --> (A --> **Prop**) --> **Prop** .

Indeed, let Γ" be the context:

Γ' [P: A --> **Prop**] [Q: A --> **Prop**] [x:A] .

Continuing from $\Gamma' \vdash A : \mathbf{Kind}$, it is readily shown that Γ'' is valid, hence:

i) $\Gamma'' \vdash x : A$,
ii) $\Gamma'' \vdash P: A \text{ --> } \mathbf{Prop}$,
iii) $\Gamma'' \vdash Px : \mathbf{Prop}$,
iv) $\Gamma''\ [y:Px]$ is valid ,
v) $\Gamma''\ [y:Px] \vdash x : A$,
vi) $\Gamma''\ [y:Px] \vdash Q: A \text{ --> } \mathbf{Prop}$,
vii) $\Gamma''\ [y:Px] \vdash Qx : \mathbf{Prop}$,
viii) $\Gamma'' \vdash (Px) \Rightarrow (Qx) : \mathbf{Prop}$,

and thus the desired typing judgment for *Subset* follows by a quantification and two abstractions.

Let us pause to realize what we have constructed so far. Fix an arbitrary $c \in ext(\Gamma')$. All values will depend on c (except $\|\mathbf{Prop}\|$, which is always the collection of quotients of equivalence relations on subsets of natural numbers). We shall often suppress c for the sake of brevity.

First note that $\| A \text{ --> } \mathbf{Prop} \|$ is simply the set of all functions from $\|A\|$ to $\|\mathbf{Prop}\|$. Thus $ext(\Gamma'')$ consists of all tuples of the form $\langle c, f, g, a\rangle$, where $a \in \|A\|$ and f and g are functions from $\|A\|$ to $\|\mathbf{Prop}\|$. Thus:

i) $\|x\|(c, f, g, a) = a$,
ii) $\|P\|(c, f, g, a) = f$,
iii) $\|Px\|(c, f, g, a) = f(a)$,
iv) $ext(\Gamma''\ [y:Px])$ consists of all tuples $\langle c, f, g, a, \tilde{n}\rangle$, where $\langle c, f, g, a\rangle \in ext(\Gamma'')$ and the equivalence class $\tilde{n}$ belongs to $f(a)$,
v) $\|x\|(c, f, g, a, \tilde{n}) = a$,
vi) $\|Q\|(c, f, g, a, \tilde{n}) = g$,
vii) $\|Qx\|(c, f, g, a, \tilde{n}) = g(a)$,
viii) $\|(Px) \Rightarrow (Qx)\|(c, f, g, a) = \Pi rec_{b \in f(a)}\ g(a)$ (that is: the set of numerical codes of partial recursive functions that map $f(a)$ to $g(a)$; modulo the required equivalence relation).

Thus $\|(x:A)\ (Px)\Rightarrow(Qx)\|\ (c,f,g)$ is obtained from $\Pi rec_{b\epsilon f(a)}\ g(a)$ by intersection over all $a \in \|A\|$, and hence $\|Subset\|(c)$ is given as the assignment:

$$\|Subset\|(c)(f)(g) \ := \ \|(x:A)\ (Px)\Rightarrow(Qx)\|\ (c,f,g)$$

for any $f, g: \|A\| \dashrightarrow \|\mathbf{Prop}\|$. (The notation in calculus of constructions here nicely coincides with the ordinary mathematical notation for functions from a given domain to a given codomain.)

Now we continue with the construction by defining the intersection of a class of predicates on A. Let *Inter* be the term:

$$[C: (A \dashrightarrow \mathbf{Prop}) \dashrightarrow \mathbf{Prop}]\ [x:A]\ (P: A \dashrightarrow \mathbf{Prop})\ (CP)\Rightarrow(Px) .$$

The reader will easily check that:

$$\Gamma' \vdash \mathit{Inter} : ((A \dashrightarrow \mathbf{Prop}) \dashrightarrow \mathbf{Prop}) \dashrightarrow A \dashrightarrow \mathbf{Prop} .$$

The value of the type is again obtained simply by reading the arrows as "the set of all functions from ... to ..." . Thus given a mapping F that takes functions from $\|A\|$ to $\|\mathbf{Prop}\|$ to the quotients in $\|\mathbf{Prop}\|$, and given a in $\|A\|$, we may obtain $\|Inter\|(c)(F)(a)$ as the quotient of the set of numerical codes of partial recursive functions that map $F(f)$ to $f(a)$ for every $f: \|A\| \dashrightarrow \|\mathbf{Prop}\|$. Two such codes e and e' are equivalent iff for every $f: \|A\| \dashrightarrow \|\mathbf{Prop}\|$, if n and m are equivalent in $F(f)$, then $\{e\}(n)$ and $\{e'\}(m)$ are equivalent in $f(a)$.

The reader will have noticed that the logical structure of this definition is exactly the one expressed in the syntax, namely the universal quantification of an implication. In fact, realizability interpretation provides a way of reading the syntax of the calculus of constructions in the ordinary mathematical way, except that proofs are read as codes of partial recursive functions. (The equivalence relations are there to account for the rules *AbsEq* and *ApplEq*.) Square brackets are read "let ... be in ..." .

We continue with the construction. Let Γ be the valid context:

$$\Gamma' \ [C_0 : (A \dashrightarrow \mathbf{Prop}) \dashrightarrow \mathbf{Prop}] \ [P_0 : A \dashrightarrow \mathbf{Prop}] \ [p_0 : (C_0 \ P_0)]$$

In this context, we shall construct a proof of the proposition that the predicate $(\mathit{Inter}\ C_0)$ is included in the predicate P_0. Let Δ be the valid context:

$$\Gamma \ [x{:}A] \ [h{:}((\mathit{Inter}\ C_0)\ x)] \ .$$

The inference rule *Beta* yields:

$$\Gamma \ [x{:}A] \ \vdash \ ((\mathrm{Inter}\ C_0)\ x) \ = \ (P{:}\ A \dashrightarrow \mathbf{Prop}) \ \ (C_0\ P) \Rightarrow (Px) \ .$$

Because this is an equality of propositions, we may use the inference rule *Equal* to obtain:

$$\Delta \ \vdash \ h : (P{:}\ A \dashrightarrow \mathbf{Prop}) \ (C_0\ P) \Rightarrow (Px) \ ,$$

hence:

$$\Delta \ \vdash \ ((h\ P_0)\ p_0) \ : \ (P_0\ x) \ .$$

Using *Beta* again, we obtain:

$$\Gamma \ \vdash \ ((\mathit{Subset}\ (\mathit{Inter}\ C_0))\ P_0) \ = \ (x{:}A) \ ((\mathit{Inter}\ C_0)\ x) \Rightarrow (P_0\ x)$$

This is also an equality of propositions, hence the rule *Equal* yields:

$$\Gamma \vdash \ [x{:}A] \ [h{:}((\mathit{Inter}\ C_0)\ x)] \ ((h\ P_0)\ p_0) \ : \ ((\mathit{Subset}\ (\mathit{Inter}\ C_0))\ P_0).$$

We conclude the discussion of this example by describing the partial recursive function that realizes the Γ-proof just constructed. This function must depend only on the construction, not on $c \in \mathrm{ext}(\Gamma')$. Let $k = \#\Gamma'$, hence $\#\Gamma = k+1$ and $\#\Delta = k+2$. $\|((h\ P_0)\ p_0)\|$ is given by the partial recursive function φ that computes $\{e\}(n)$ from input $\langle i_1, \ldots, i_k, n, e\rangle$. Our Γ-proof is then realized by the partial recursive function that computes the s-m-n code of φ for any given $i_1, \ldots, i_k$, and n. ∎

The calculus interpreted here is basically the *pure* calculus of constructions considered in Huet [87], but our interpretation also extends to various stronger calculi considered there. Such extensions are facilitated by considering the interpretation described here as an interpretation in a fragment of the (Recursive) Realizability Universe rather than in the universe of ordinary sets (see section 4 above). We have interpreted (x:A)B as products given intrinsically in the Realizability Universe. This point of view makes it plausible to consider notions of valid contexts that are much more extensive than in the pure calculus.

We also observe that the realizability interpretation itself may be formalized in an appropriate fragment of intuitionistic set theory (depending on a particular version of the calculus), and thus it may be used in obtaining information about the logical power of calculus of constructions. This information may be formulated by means of numerical functions representable in the calculus. Formalizing the recursive realizability interpretation of the *pure calculus of constructions* in higher-order arithmetic yields the result that the numerical functions representable in this calculus (as proofs of the proposition $Nat \Rightarrow Nat$) are exactly the numerical functions representable in Girard's higher-order polymorphic lambda calculus F^{ω} , to wit, the recursive functions provably total in higher-order arithmetic. Stronger versions of the calculus correspond in this manner to stronger systems of set theory.

REFERENCES

Amadio, R., Bruce, K.B., Longo, G. [86] The finitary projection model for second-order lambda calculus and solutions to higher-order domain equations. *Proc 1st IEEE Symposium on Logic in Computer Science*, Cambridge, Mass., June 1986.

Bainbridge, E.S., Freyd, P.J., Scedrov, A., Scott, P.J. [87] Functorial polymorphism. In: *Logical Foundations of Functional Programming, Proceedings University of Texas Programming Institute, Austin, Texas, June 1987*, ed. by G. Huet, to appear.

Barendregt, H. [84] *The lambda calculus. Its syntax and semantics.* Revised edition, North-Holland, Amsterdam, 1984.

Barendregt, H. [8x] Lambda calculi with types. In: *Handbook of logic in computer science*, ed. by S. Abramsky et al., Oxford Univ. Press, to appear.

Barnes, J.G.P. [81] *Programming in Ada.* Addison-Wesley, 1981.

Breazu-Tannen, V. [88] Combining algebra and higher types. *Proc. 3rd IEEE Symposium on Logic in Computer Science*, Edinburgh, Scotland July 1988.

Breazu-Tannen, V., Buneman, O.P., Gunter, C.A. [88] Typed functional programming for rapid development of reliable software. *Proc. ACM Symposium on Productivity: Prospects, Progress, and Payoff.* Washington, D.C. chapter of the ACM, June 1988.

Breazu-Tannen, V., Coquand, T. [87] Extensional models for polymorphism. *Proc. TAPSOFT '87 - CFLP, Pisa.* Springer LNCS 250. Expanded version to appear in *Theor. Comp. Science.*

Breazu-Tannen, V., Meyer, A.R. [87] Computable values can be classical. *Proc. 14th Annual ACM Symposium on Principles of Programming Languages*, Munich, West Germany, January 1987.

Bruce, K.B., Longo, G. [88] A modest model of records, inheritance, and bounded quantification. *Proc. 3rd IEEE Symposium on Logic in Computer Science*, Edinburgh, Scotland, July 1987.

Bruce, K.B., Meyer, A.R., Mitchell, J.C. [8x], The semantics of second-order lambda calculus. *Information and Computation*, to appear.

Carboni, A., Freyd, P., Scedrov, A. [88] A categorical approach to realizability and polymorphic types. *Proc. 3rd ACM Workshop on the Mathematical Foundations of the Programming Language Semantics, New Orleans, April, 1987*, ed. by M. Main et al., Springer LNCS 298, 1988, pp. 23-42.

Cardelli, L. [88] Time for a new language. *Preprint*, April 1988.

Constable, R.L., et al. [86] *Implementing mathematics with the NUPRL proof development system.* Prentice Hall, 1986.

Coquand, T. [85] *Une théorie des constructions.* Thèse de troisème cycle, Université Paris VII.

Coquand, T., Huet, G. [85] *Constructions: a higher-order proof system for mechanizing mathematics.* Proc. EUROCAL '85, Springer LNCS 203, pp. 151-184.

Coquand, T., Huet, G. [87] Concepts mathématiques et informatiques formalisés dans le calcul des constructions. In: *Logic Colloquium 85* (ed. by The Paris Logic Group), North-Holland, Amsterdam, 1987.

Coquand, T., Gunter, C.A., Winskel, G. [88a] dI-domains as a model of polymorphism. *Proc. 3rd ACM Workshop on the Mathematical Foundations of the Programming Language Semantics, New Orleans, April 1987*, ed. by M. Main et al., Springer LNCS 298, 1988, pp. 344-363.

Coquand, T., Gunter, C.A., Winskel, G. [88b] Domain theoretic models for polymorphism. *Information and Computation*, to appear.

Cousineau, G. [87] CAML. *Lectures at the University of Texas Programming Institute on the Logical Foundations of Functional Programming*, Austin, Texas, June 1987.

Cousineau, G., Curien, P.L., Robinet, B. (eds.) [86], *"Combinators and Functional Programming Languages".* Springer LNCS 242.

Curien, P.L. [86], *Categorical combinators, sequential algorithms, and functional programming.* Research notes in theoretical computer science, Pitman, 1986.

de Bruijn, N.G. [72] Lambda calculus notation with nameless dummies, a tool for automatic formula manipulation. *Indagationes Math.* 34 (1972) pp. 381-392.

Freyd, P., Scedrov, A. [87] Some semantic aspects of polymorphic lambda calculus. *Proc. 2nd IEEE Symposium on Logic in Computer Science*, Ithaca, NY, 1987, pp. 315-319.

Freyd, P. J., Girard, J. Y., Scedrov, A., Scott, P. J. [88] Semantic parametricity in polymorphic lambda calculus. *Proc. 3rd IEEE Symposium on Logic in Computer Science*, Edinburgh, Scotland, July 1988.

Giannini, P., Honsell, F., Ronchi Della Rocca, S. [87] A strongly normalizing term having no type in the system F (second-order λ-calculus). *Rapporto Interno, Dipartimento di Informatica, Università di Torino*, 1987.

Giannini, P., Ronchi Della Rocca, S. [88] Characterization of typing in polymorphic type discipline. *Proc. 3rd IEEE Symposium on Logic in Computer Science*, Edinburgh, Scotland, July 1988.

Girard, J. Y. [71], Une extension de l'interprétation de Gödel ... In: *Second Scandinavian Logic Symposium, 1970*, ed. by J. E. Fenstad, North-Holland, Amsterdam, 1971.

Girard, J. Y. [72] *Interprétation fonctionelle et élimination des coupures de l'arithmétique d'ordre supérieur.* These de Doctorat d'Etat, Université Paris VII, 1972.

Girard, J. Y. [86] The system F of variable types, fifteen years later. *Theor. Comp. Science* 45 (1986) pp. 159-192.

Girard, J. Y. [87a] Linear logic. *Theor. Comp. Science* 50 (1987) pp. 1-102.

Girard, J. Y. [87b] *Proof theory and logical complexity.* Studies in proof theory, Bibliopolis, Napoli, 1987.

Gordon, M. J. C., Milner, R., Wadsworth, C. [79] *Edinburgh LCF.* Springer LNCS 78 , 1979.

Gunter, C. A., Scott, D. S. [88] Semantic domains. In: *Handbook of Theoretical Computer Science*, ed. by J. van Leeuwen, North-Holland, Amsterdam, to appear.

Hindley, J.R., Seldin, J.P. [86] *Introduction to combinators and lambda calculus.* Cambridge University Press, 1986.

Howard, W.A. [69] The formulae-as-types notion of construction. *Unpublished manuscript*, 1969. Reprinted in: *To H.B. Curry: Essays on combinatory logic, lambda calculus, and formalism.* J.P. Seldin and J.R. Hindley, eds., Academic Press, 1980.

Huet, G. [80] Confluent reductions: abstract properties and applications to term rewriting systems. *J.A.C.M.* 27 (1980) pp. 797-821.

Huet, G. [86], Deduction and computation. In: *Fundamentals in Artificial Intelligence*, eds. W. Bibel and P. Jorrand, Springer LNCS 232, 1986.

Huet, G. [87] A uniform approach to type theory. In: *Logical Foundations of Functional Programming, Proceedings University of Texas Programming Institute, Austin, Texas, June 1987*, ed. by G. Huet, to appear.

Huet, G., and Lévy, J.J. [79] Call by need computations in non-ambiguous linear term rewriting systems. *Rapport Laboria* 359, *IRIA*, August 1979.

Hyland, J.M.E. [87] A small complete category. *Preprint*, 1987.

Hyland, J.M.E., Robinson, E.P., Rosolini, G. [87] The discrete objects in the Effective Topos. *Preprint, 1987.*

Hyland, J.M.E., Pitts, A. [87] The theory of constructions: categorical semantics and topos-theoretic models. In: Categories in Computer Science and Logic, Proceedings Amer. Math. Soc. Research Conference, Boulder, Colorado, June 1987, ed. by J.W. Gray and A. Scedrov, to appear.

Klop, J.W. [80] *Combinatory reduction systems.* Ph.D. Dissertation, Mathematisch Centrum Amsterdam, 1980.

Krivine, J.L. [87] Un algorithme non typable dans le système F. *Compt. Rend. Acad. Sci. Paris, Ser. I, Math.* 304 No. 5 (1987) pp. 123-126.

Lamarche, F. [87] A model for the theory of constructions. In: *Categories in Computer Science and Logic, Proceedings Amer. Math. Soc. Research Conference, Boulder, Colorado, June 1987*, ed. by J.W. Gray and A. Scedrov, to appear.

Lambek, J., Scott, P.J. [86] *Introduction to higher-order categorical logic.* Cambridge University Press.

Leivant, D. [83] Reasoning about functional programs and complexity classes associated with type disciplines. *24th Annual IEEE Symposium on Foundations of Computer Science*, 1983.

Lévy, J.J. [78] *Réductions correctes et optimales dans le λ-calcul.* Thèse d'Etat, Université de Paris VII, 1978.

Liskov, B. et al. [81] *Clu reference manual.* Springer LNCS 114, 1981.

Longo, G., Moggi, E. [87] Constructive natural deduction and its "modest" interpretation. *Workshop on semantics of natural and computer languages, Stanford, March 1987*, ed. by J. Meseguer et al., M.I.T. Press, to appear.

MacQueen, D. [85] Modules for Standard ML. *Polymorphism Newsletter* 2(2), 1985.

Martin-Löf, P. [84] *Intuitionistic type theory.* Studies in proof theory, Bibliopolis, Napoli, 1984.

McCracken, N. [79] *An investigation of a programming language with polymorphic type structure.* Ph.D. Dissertation, Syracuse University, 1979.

Meseguer, J. [88] Relating models of polymorphism. *Technical note* SRI-CSL-TN88-1, SRI International, June 1988.

Meyer, A.R. [88] Invited lecture at the 3rd IEEE Symposium on Logic in Computer Science, Edinburgh, July, 1988.

Milner, R. [84] A proposal for standard ML. In: *ACM Symposium on LISP and Functional Programming*, 1984, pp. 184-197.

Mitchell, J.C. [84] Type inference and type containment. In: *Symp. on Semantics of Data Types*. Springer LNCS 173, 1984, pp. 257-278, revised version to appear in *Information and Computation*.

Mitchell, J.C. [86] A type-inference approach to reduction properties and semantics of polymorphic expressions. In: *Proc. 1986 ACM Symposium on Lisp and Functional Programming*, pp. 308-319.

Mitchell, J.C., Meyer, A.R. [85] Second-order logical relations. In: *Logics of Programs*, ed. by. R. Parikh, Springer LNCS 193, 1985, pp. 225-236.

Mitchell, J.C., Plotkin, G.D. [85] Abstract types have existential types. In: *Proc. 12th ACM Symposium on Principles of Programming Languages*, January 1985, pp. 37-51.

Moggi, E. [88] Dissertation, Edinburgh University. *In preparation.*

Mohring, C. [86] Algorithm development in the calculus of constructions. *Proc. 1st IEEE Symposium on Logic in Computer Science*, Cambridge, Mass., 1986, pp. 84-91.

Pitts, A. [87] Polymorphism is set-theoretic, constructively. *Symposium on Category Theory and Computer Science*, Springer LNCS 283, 1987.

Plotkin, G.D. [85] Denotational semantics with partial functions. *Lecture notes, CSLI Summer School*, Stanford, 1985.

Pottinger, G. [87] Strong normalization for terms of the theory of constructions. *Preprint*, February 1987.

Prawitz, D. [65] *Natural deduction*. Almquist and Wiksell, Stockholm, 1965.

Reynolds, J.C. [74] Towards a theory of type structure. *Springer LNCS* 19 , 1974, pp. 408-425.

Reynolds, J.C. [83] Types, abstraction, and parametric polymorphism. In: *Information Processing '83*, ed. by R.E.A. Mason. North-Holland, Amsterdam, pp. 513-523.

Reynolds, J.C. [84] Polymorphism is not set-theoretic. *Symposium on Semantics of Data Types*, ed. by Kahn et al., Springer LNCS 173 , 1984.

Reynolds, J.C., Plotkin, G.D. [87] On functors expressible in the polymorphic typed lambda calculus. Preliminary report in: *Logical Foundations of Functional Programming, Proceedings University of Texas Programming Institute, Austin, Texas, June 1987*, ed. by G. Huet, to appear. Extended version to appear in *Information and Computation.*

Rosolini, G. [86] About Modest Sets. *Preprint*, 1986.

Scedrov, A. [87a] Recursive realizability interpretation of calculus of constructions. In: *Logical Foundations of Functional Programming, Proceedings University of Texas Programming Institute, Austin, Texas, June 1987*, ed. by G. Huet, to appear.

Scedrov, A. [87b] Normalization revisited. In: *Categories in Computer Science and Logic, Proceedings Amer. Math. Soc. Research Conference, Boulder, Colorado, June 1987*, ed. by J.W. Gray and A. Scedrov, to appear.

Scott, D.S. [72] Continuous lattices. In: *Toposes, Algebraic Geometry and Logic*, ed. by F.W. Lawvere, Springer LNM 274, 1972, pp. 97-136.

Scott, D.S. [76] Data types as lattices. *SIAM J. of Computing* 5 (1976) pp. 522-587.

Scott, D.S. [82] Domains for denotational semantics. *ICALP '82*, Springer LNCS 140 .

Scott, D. S. [87] Realizability and domain theory. *Lecture at the Amer. Math. Soc. Research Conference on Categories in Computer Science and Logic*, Boulder, Colorado, June 1987.

Seely, R. A. G. [87a] Categorical semantics for higher-order polymorphic lambda calculus. *J. Symbolic Logic* 52 (1987) pp. 969-989.

Seely, R. A. G. [87b] Modelling computations: a 2-categorical framework. *2nd IEEE Symposium on Logic in Computer Science*, Ithaca, NY, pp. 65-71.

Seldin, J. [87] Theory of MATHESIS. *Technical Report, Odyssey Research Associates, Inc.*, March 1987.

Smyth, M. B., Plotkin, G. D. [82] The category-theoretic solution of recursive domain equations. *SIAM J. of Computing* 11 (1982) pp. 761-783.

Statman, R. [81] Number theoretic functions computable by polymorphic programs. *22nd Annual IEEE Symposium on Foundations of Computer Science*, 1981.

Stenlud, S. [72] *Combinators, λ-terms, and proof theory.* Reidel, 1972.

Strachey, C. [67] Fundamental concepts in programming languages. Lecture Notes, *International Summer School in Computer Programming*, Copenhagen, August 1967.

Tait, W. W. [75] A realizability interpretation of the theory of species. In: Springer LNM 453 , 1975, pp. 240-251.

Takeuti, G. [87] *Proof theory.* Second edition, North-Holland, Amsterdam, 1987.

Turner, D. A. [85] Miranda: a non-strict functional language with polymorphic types. In: *Functional Pragramming Languages and Computer Architecture*, ed. by J. P. Jouannaud, Springer LNCS 201 , 1985, pp. 1-16.

C.I.M.E. Session on "Logic and Computer Sciences"

List of Participants

V.M. ABRUSCI, Viale dei Mille 40, 50131 Firenze

G. AGUZZI, Dipartimento di Sistemi e Informatica, Via S. Marta 3, 50139 Firenze

C. ALVAREZ FAURA, Facultat d'Informatica, c/ Paul Gargallo 5, 08028 Barcelona

F. ARZARELLO, Dipartimento di Matematica, Via C. Alberto 10, 10123 Torino

E. BALLO, Via M. Dal Re 24, 20156 Milano

F. BARBANERA, Via Oriani 10, 04100 Latina

L. BERNIS, 65 rue Maurice Riposch, F-75014 Paris

P. BERTAINA, Dipartimento di Informatica, Corso Svizzera 185, 10149 Torino

F.A. BRUNACCI, Istituto M.A.S.E.S., Via Montebello 7, 50123 Firenze

D. BRUSCHI, Dipartimento di Scienze dell'Informazione, Via M. Da Brescia 9, 20133 Milano

A. CANTINI, Dipartimento di Filosofia, Via Bolognese 52, 50139 Firenze

D. CANTONE, Via E. D'Angiò 46 E/2, 95125 Catania

A. CARBONE, Dipartimento di Matematica, Via del Capitano 15, 53100 Siena

F. CARDONE, Via Boston 108/33, 10137 Torino

L.E. CASTILLO HERN, 80 South Bridge, Edinburgh, EH1 1HN

M. CHIARI, Via Chiarugi 12, 50136 Firenze

J. CHRISTENSEN, Mathematical Institute, The Technical University of Denmark, Building 303, 2800 Lyngby

A. CORRADINI, Dipartimento di Informatica, Corso Italia 40, 56100 Pisa

G. CRISCUOLO, Dipartimento di Scienze Fisiche, Mostra d'Oltremare, Pad. 19, 80125 Napoli

V. CUTELLO, Dipartimento di Matematica, Viale A. Doria 6, 95125 Catania

G.B. DEMO, Dipartimento di Informatica, Corso Svizzera 185, 10149 Torino

F. DORINI, Dipartimento di Informatica e Sistemistica, Università di Roma

V. DRAPERI, Via Cesare Battisti 15, 10123 Torino

G. EPSTEIN, Department of Computer Science, University of North Carolina at Charlotte, Charlotte, NC 28223

M. FALASCHI, Dipartimento di Informatica, Corso Italia 40, 56100 Pisa

G.L. FERRARI, Dipartimento di Informatica, Corso Italia 40, 56100 Pisa

P.L. FERRARI, Dipartimento di Matematica, Via L.B. Alberti 4, 16132 Genova

A. FERRO, via Pietro Carrera 2, 95123 Catania

C. FRANCIA, Via A.V.I.S. 2, 10048 Vinovo (Torino)

C. FURLANELLO, IRST, Loc. Pantè di Povo, 38100 Trento

P. GENTILINI, Piazza Leopardi 16/2, 16145 Genova

S. GHILARDI, Via Belestra 5, 24100 Bergamo

L. GIORDANO, Dipartimento di Informatica, Corso Svizzera 185, 10149 Torino

E. GIOVANNETTI, Dipartimento di Informatica, Corso Svizzera 185, 10149 Torino

S. GNESI, IEI-CNR, Via S. Maria 46, 56100 Pisa

J.-M. GRANDMONT, Université de l'Etat, Faculté des Sciences, 15 Avenue Mistriau, 7000 Mons

Y. HARTMANIS, Department of Computer Science, Cornell University, Ithaca, NY 14853

D. HASKELL, Mathematisches Institut, Beringstr. 4, 5300 Bonn 1

S. HOMER, Department of Computer Science, Boston University, Boston, Mass. 02215

P. INVERARDI, IEI-CNR, Via S. Maria 46, 56100 Pisa

M. KAPETANOVIC, Matematicki Institut, Knez Mihailova 35, 11000 Beograd

G. KOLETSOS, 41-43 Ioylianoy St., Athens 104 33

J. LAMBEK, Mathematics Department, McGill University, Montréal, QUe. H3A 2K6

D. LATCH, Department of Computer Science, Brooklyn College Cuny, Brooklyn, NY 11210

D. LATELLA, CNR-CNUCE, Via S. Maria 36, 56100 Pisa

J. LIPTON, Department of Mathematics, White Hall, Cornell University, Ithaca, NY 14853

E. LOCURATOLO, IEI-CNR, Via S. Maria 46, 56100 Pisa

G. LOLLI, Dipartimento di Informatica, Corso Svizzera 185, 10149 Torino

A. LOZANO BOJADOS, Facultat d'Informàtica, Pau Gargallo 5, 08028 Barcelona

S. MALECKI, 64 rue vergniaud, 75013 Paris

P. MANGANI, Istituto Matematico U. Dini, Viale Morgagni 67/A, 50134 Firenze

C. MANGIONE, via G. Giusti 3, 20154 Milano

E. MARCHIORI, via Ponte Piana 37, 30170 Mestre (Venezia)

A. MARCJA, Dipartimento di Matematica, 38050 Povo (Trento)

G. MARONGIU, Via G. del Piani dei Carpini 96/B, 50127 Firenze

N. MARTI-OLIET, Departamento de Informatica y Automatica, Facultad de Matematicas, Universidad Complutense de Madrid, 28040 Madrid

S. MARTINI, Dipartimento di Informatica, Corso Italia 40, 56100 Pisa

S. MATTHEWS, University of Edinburgh, Department of Artificial Intelligence, 80 South Bridge, Edinburgh EH1 1HN

G.-C. MELONI, Dipartimento di Matematica, Via C. Saldini 50, 20133 Milano

C. MIROLO, Dipartimento di Matematica e informatica, Via Zanon 6, 33100 Udine

D. MUSTO, IEI-CNR, Via S. Maria 46, 56100 Pisa

A. NERODE, Department of Mathematics, Cornell University, Ithaca, NY 14853

P. ODIFREDDI, Dipartimento di Informatica, Corso Svizzera 185, 10149 Torino

N. OUAKRIM, 9 rue d'Houdain, 7000 Mons

G. PANTI, Via Cappuccini 128, 53100 Siena

F. PARLAMENTO, Dipartimento di Matematica e Informatica, Via Zanon 6, 33100 Udine

J.P. PEDERSEN, Mathematical Institute, The Technical University of Denmark, Building 303, DK-2800 Lyngby

L. PERO, Via Tertulliano 41, 20137 Milano

R. PLATEK, Odissey Research Association, 1283 Trumansbury, Ithaca, NY 14850

A. POLICRITI, Corso Risorgimento 6, 13051 Biella

T. PULS, The Technical University of Denmark, Building 345V.274, DK-2800 Lyngby

C. RODINE, 21 rue des Cordelières, 75013 Paris

G. ROSOLINI, Dipartimento di Matematica, Via Università 12, 43100 Parma

R. RUGGERI CANNATA, Dipartimento di Matematica, Viale A. Doria 6, 95125 Catania

G. SACHS, Department of Mathematics, Harvard University, Cambridge, Mass. 02138

A. SCEDROV, Department of Mathematics, Univ. of Pennsylvania, Philadelphia, PA 19104

M. SCHERF, Dipartiemnto di Informatica e Sistemistica, Università di Roma

F. SEBASTIANI, IEI-CNR, Via S. Maria 46, 56100 Pisa

R. SIGAL, Dipartimento di Matematica, Viale A. Doria 6, 95125 Catania

U. SOLITRO, Dipartimento di Scienze dell'Informazione, Via Moretto da Brescia 9, 20133 Milano

L. TERRACINI, Corso Dante 118, 10126 Torino

C. TOFFALORI, Istituto Matematico U. Dini, Viale Morgagni 67/A, 50134 Firenze

S. TULIPANI, Dipartimento di Matematica e Fisica, Università di Camerino, 62032 Camerino (Macerata)

A. URSINI, Dipartimento di Matematica, Via del Capitano 15, 53100 Siena

S. VALENTINI, Via G. Malaspina 9, 35124 padova

B. VENNERI, Dipartimento di Informatica, Corso Svizzera 185, 10149 Torino

T. WALSH, University of Edinburgh, Department of Artificial Intelligence, 80 South Bridge, Edinburgh EH1 1HN

W. WERNECKE, IBM Scientific Center, Wilckenstr. 1a, 6900 Heidelberg

L. ZAVATTARO, Via Sempione 214, 10154 Torino

FONDAZIONE C.I.M.E.
CENTRO INTERNAZIONALE MATEMATICO ESTIVO
INTERNATIONAL MATHEMATICAL SUMMER CENTER

"Methods of nonconvex analysis"

is the subject of the First 1989 C.I.M.E. Session.

The Session, sponsored by the Consiglio Nazionale delle Ricerche and the Ministero della Pubblica Istruzione, will take place under the scientific direction of Prof. ARRIGO CELLINA (S.I.S.S.A., Trieste), at Villa "Monastero", Varenna, Lake of Como, Italy, ***from June 15 to June 23, 1989.***

Courses

a) ***The ε Variational Principle revised.*** (4 lectures in English).
Prof. Ivar EKELAND (Université Paris Dauphine, France).

Outline

In these lectures, we shall investigate some of the progress which intervened since the 1972 version of the ε variational principe. We shall describe the higher order version due to Frankowska (1987) and the smooth version of Borwein and Preiss (1988). We shall go into the theory of mountain pass points, that is the Ambrosetti-Rabinowitz existence results and the characterization of Hofer and Pucci-Serrin, using the theorem of Ghoussoub (1988). If time permits we may give some applications to dynamical systems or to nonlinear elasticity.

b) ***Problems of the Calculus of Variations involving Nonconvex Integrals*** (4 lectures in English).
Prof. Paolo MARCELLINI (Università di Firenze, Italia)

Outline

We will consider integrals of the calculus of variations ot the type

$$F(u) = \int_{\Omega} f(x,u,Du)\,dx\;;$$

Ω is a bounded open set of R^n and u is defined in Ω and has its values in R^N; Du is the gradient of u. The function f $= f(x,u,v)$ is not necessarily convex with respect to v.

Minimizers of F in a given class of functions may or may not exist. We will consider separately the four cases: (1) $n = 1$, $N = 1$; (2) $n = 1$, $N > 1$, (3) $n > 1$, $N = 1$; (4) $n > 1$, $N > 1$.

In each of them we will present sufficient conditions for the existence of minimizers and some problems not yet solved. In particular, in the case (4), we will describe the quasi-convexity condition by Morrey and its applications to nonlinear elasticity.

c) ***Lower Critical Points and Curves of Maximal Slope in Problems with Lack of Convexity.*** (4 lectures in English)
Prof. Antonio MARINO (Università di Pisa, Italia)

Ouline

"Lower critical points" and "curves of maximal slope" are considered for suitable classes of lower semicontinous functionals, non necessarily convex, possibly in the case of nonsmooth and nonconvex constraints. In this framework, semilinear equations and variational inequalities of elliptic (related to lower critical points) and parabolic (related to curves of maximal slope) type, are solved. In the case of variational inaqualities some nonconvex constraint condi tions are considered.

The lack of convexity gives rise to a multiplicity of solutions for some variational inequalities of elliptic type. To prove these results the flow of the corresponding parabolic variational inequality is used.

References

1) M. Degiovanni and A. Marino, Eigenvalue Problems for some Nonlinear Elliptic Varational Inequalities and related Evolution Equations, Res. Notes in Math. 125, Pitman, 1986.
2) A. Marino, Evolution Equations and Multiplicity of Critical Points with respect to an Obstacle. Res. Notes in Math. 148, Pitman, 1987.
3) A. Marino, C. Saccon and M. Tosques, Curves of Maximal Slope and Parabolic Variational Inequalities on Nonconvex Constraints, Ann. Scuola Norm. Sup. (to appear).
4) M. Degiovanni, A. Marino and M. Tosques, Evolution Equations with Lack of Convexity, Nonlinear Analysis, T.M.A. 9, 1985.

d) ***The Liapunov Theorem: its extensions and applications.*** (4 lectures in English)
Prof. Czeslaw OLECH (P.A.N., Warszawa, Poland)

Contents

The Liapunov theorem on the range of vector valued monatomic measures. Integrations of decomposable subsets of L^1 (Auman integral).

Extreme point of a decomposable set and an analogue of Carathéodory theorem.

The unbounded case. The extension to the infinite dimensional case.

Applications to: the theory of optimal control and the calculus of variations (existence of optimal solutions, conditions for optimality, bang-bang principle, lower semicontinuity of integral functionals); to differential inclusions (continuous selections) to decomposable set-valued maps and existence of solutions of differential inclusions in the nonconvex case) and further applications.

References

1) A.A. Lyapunov, Sur les fonctions vécteurs completement additives, Bull. Acad. Sci. U.R.S.S., sér. Math. 4, 1940, 465-478.
2) H. Hermes and J.P. LaSalle, Functional Analysis and Time Optimal Control, Academic Press 1969.
3) C. Castaing and M. Valadier, Convex Analysis and Measurable Multifunctions, Lecture Notes in Math. 580, Springer 1977.
4) J.P. Aubin and A. Cellina, Differential Inclusions, Springer 1984.

e) ***Differential Inclusions: the Category Method.*** (4 lectures in English).
Prof. Giulio PIANIGIANI (Università di Siena, Italia)

Contents

Convex and nonconvex differential inclusions in R^n. Classical and absolutely continuous solutions. Selections, Filippov example. Filippov existence theorem; the method of Antosiewicz-Cellina.

The solution sets of (1) $x' \in F(t,x)$, $x(t_0) = x_0$ and (2) $x' \in coF(t,x)$, $x(t_0) = x_0$, in finite and infinite dimensional spaces, Plis example.

Relations between the solution sets of (1) and (2) via the category approach.

Extreme points: the Choquet function. Extremal solutions and existence theorems.

References

1) J.P. Aubin and A. Cellina, Differential Inclusions, Springer 1984.
2) E.M. Alfsen, Compact Convex Sets and Boundary Integrals, Springer 1971.
3) C. Castaing and M. Valadier, Convex Analysis and Measurable Multifunctions, Lecture Notes in Math. 580, Springer 1977.

f) ***Nonsmooth Analysis and Parametric Optimization.*** (4 lectures in English)
Prof. Tyrrell R. ROCKAFELLAR (University of Washington, Seattle, USA)

Contents

First and second order derivatives defined by epi-convergence of difference quotients. Subgradients.
First and second order optimality conditions, necessary and sufficient.
Graphical differentiation of set-valued mappings, including subgradient mappings.
Generalized derivatives in analyzing the dependence of optimal solutions on parameters.
Auxiliary properties for computing such derivatives.

g) ***Young Measures.*** (4 lectures in English)
Prof. Michel VALADIER (U.S.T.L., Montpellier, France)

Contents

Young measures: locally compact and general cases.
Tightness.
Applications to sequences of integrable functions.

References

1) Berliocchi and Lasry, Integrands normales et mésures parametrées en calcul des variations, Bull. Soc. Math. France 101, (1973), pp. 129-184.
2) Balder, A General Approach to Lower Semicontinuity and Lower Closure in Optimal Control Theory. SIAM J. Control and Opt. 22, (1984), pp. 570-598.
3) Ball, A Version of the Fundamental Theorem for Young Measures, to appear.

Seminars

A number of seminars and special lectures will be offered during the Session.

FONDAZIONE C.I.M.E.
CENTRO INTERNAZIONALE MATEMATICO ESTIVO
INTERNATIONAL MATHEMATICAL SUMMER CENTER

"Microlocal Analysis and Applications"

is the subject of the Second 1989 C.I.M.E. Session.

The Session, sponsored by the Consiglio Nazionale delle Ricerche and the Ministero della Pubblica Istruzione, will take place under the scientific direction of Prof. LAMBERTO CATTABRIGA (Università di Bologna), and Prof. LUIGI RODINO (Università di Torino) at Villa «La Querceta», Montecatini (Pistoia), Italy, ***from July 3 to July 11, 1989.***

Courses

a) ***Microlocal Analysis in Nonlinear Partial Differential Equations.*** (6 lectures in English).
Prof. Jean-Michel BONY (Ecole Polytechnique, Palaiseau, France).

Contents

— Paradifferential Calculus
— Second Microlocalization
— Microlocalizations of higher order
— Applications to propagation and interaction of singularities for solutions of nonlinear P.D.E.

About the literature related to the course

Some knowledge of the Weyl calculus (section 18.5 in L. Hörmander, "The Analysis of Linear P.D.E.") and of Littlewood-Paley theory [R. Coifman and Y. Meyer "Au delà des Opérateurs Pseudo-Différentiels", Astérisque 57 (1978)] will be useful but not necessary. One can read also the survey: J.-M. Bony, Proc. Int. Cong. Math. Warszawa (1983), 1133-1146.

b) ***Parabolic Pseudo-Differential Boundary Problems and Applications.*** (6 lectures in English).
Prof. Gerd GRUBB (Købenavns Universitets, Denmark)

Outline

The theory of parabolic diffyrential equations has evolved from the original direct calculations of the heat equations, through functional analysis methods for more general equations (in particular semigroup theory), to refined techniques in the framework of pseudo-differential operator theory. Recently also non-local problems have been included, allowing integral operator in the interior equation or in the boundary condition, of a pseudo-differential nature.
Besides being of interest in itself, the theory of parabolic pseudo-differential boundary value problems is an efficient tool to handle various differential operator problems, e. g. cases where a degeneracy of the parabolicity can be eliminated by composition with pseudo-differential operators (examples: Navier-Stokes problem, and singular perturbation problems), or cases where other manipulations make the problem non-local (example: boundary feed-back in control theory for parabolic problems).
The theory we shall present, is a development of the Vishik-Eskin-Boutet de Monvel theory of elliptic pseudo-differential operators, as presented e. g. in [N], [H], Ch 18.1 or [T]. Besides this, it will be advantageous to have some

f) ***Nonsmooth Analysis and Parametric Optimization.*** (4 lectures in English)
Prof. Tyrrell R. ROCKAFELLAR (University of Washington, Seattle, USA)

Contents

First and second order derivatives defined by epi-convergence of difference quotients. Subgradients.
First and second order optimality conditions, necessary and sufficient.
Graphical differentiation of set-valued mappings, including subgradient mappings.
Generalized derivatives in analyzing the dependence of optimal solutions on parameters.
Auxiliary properties for computing such derivatives.

g) ***Young Measures.*** (4 lectures in English)
Prof. Michel VALADIER (U.S.T.L., Montpellier, France)

Contents

Young measures: locally compact and general cases.
Tightness.
Applications to sequences of integrable functions.

References

1) Berliocchi and Lasry, Integrands normales et mésures parametrées en calcul des variations, Bull. Soc. Math. France 101, (1973), pp. 129-184.
2) Balder, A General Approach to Lower Semicontinuity and Lower Closure in Optimal Control Theory, SIAM J. Control and Opt. 22, (1984), pp. 570-598.
3) Ball, A Version of the Fundamental Theorem for Young Measures, to appear.

Seminars

A number of seminars and special lectures will be offered during the Session.

Applications

Those who wish to attend the Session should fill in an application form and mail it to the Director of the Fondazione C.I.M.E. at the address below, ***not later than*** May 15, 1989.

An important consideration in the acceptance of applications is the scientific relevance of the Session to the field of interest of the applicant.

Applicants are requested, therefore, to submit, along with their application, a scientific curriculum and a letter of recommendation.

Participation will only be allowed to persons who have applied in due time and have had their application accepted.

Attendance

No registration fee is requested.
Lectures will be held at the Villa "Monastero" in Varenna, Lake of Como, Italy, on June 15, 16, 17, 18, 19, 20, 21, 22, 23.
Participants are requested to register at the Villa "Monastero" on June 14, 1989.

LIST OF C.I.M.E. SEMINARS

		Publisher
1954 -	1. Analisi funzionale	C.I.M.E.
	2. Quadratura delle superficie e questioni connesse	"
	3. Equazioni differenziali non lineari	"
1955 -	4. Teorema di Riemann-Roch e questioni connesse	"
	5. Teoria dei numeri	"
	6. Topologia	"
	7. Teorie non linearizzate in elasticità, idrodinamica,aerodinamica	"
	8. Geometria proiettivo-differenziale	"
1956 -	9. Equazioni alle derivate parziali a caratteristiche reali	"
	10. Propagazione delle onde elettromagnetiche	"
	11. Teoria della funzioni di più variabili complesse e delle funzioni automorfe	"
1957 -	12. Geometria aritmetica e algebrica (2 vol.)	"
	13. Integrali singolari e questioni connesse	"
	14. Teoria della turbolenza (2 vol.)	"
1958 -	15. Vedute e problemi attuali in relatività generale	"
	16. Problemi di geometria differenziale in grande	"
	17. Il principio di minimo e le sue applicazioni alle equazioni funzionali	"
1959 -	18. Induzione e statistica	"
	19. Teoria algebrica dei meccanismi automatici (2 vol.)	"
	20. Gruppi, anelli di Lie e teoria della coomologia	"
1960 -	21. Sistemi dinamici e teoremi ergodici	"
	22. Forme differenziali e loro integrali	"
1961 -	23. Geometria del calcolo delle variazioni (2 vol.)	"
	24. Teoria delle distribuzioni	"
	25. Onde superficiali	"
1962 -	26. Topologia differenziale	"
	27. Autovalori e autosoluzioni	"
	28. Magnetofluidodinamica	"

1963 - 29. Equazioni differenziali astratte C.I.M.E.
30. Funzioni e varietà complesse "
31. Proprietà di media e teoremi di confronto in Fisica Matematica "

1964 - 32. Relatività generale "
33. Dinamica dei gas rarefatti "
34. Alcune questioni di analisi numerica "
35. Equazioni differenziali non lineari "

1965 - 36. Non-linear continuum theories "
37. Some aspects of ring theory "
38. Mathematical optimization in economics "

1966 - 39. Calculus of variations Ed. Cremonese, Firenze
40. Economia matematica "
41. Classi caratteristiche e questioni connesse "
42. Some aspects of diffusion theory "

1967 - 43. Modern questions of celestial mechanics "
44. Numerical analysis of partial differential equations "
45. Geometry of homogeneous bounded domains "

1968 - 46. Controllability and observability "
47. Pseudo-differential operators "
48. Aspects of mathematical logic "

1969 - 49. Potential theory "
50. Non-linear continuum theories in mechanics and physics and their applications "
51. Questions of algebraic varieties "

1970 - 52. Relativistic fluid dynamics "
53. Theory of group representations and Fourier analysis "
54. Functional equations and inequalities "
55. Problems in non-linear analysis "

1971 - 56. Stereodynamics "
57. Constructive aspects of functional analysis (2 vol.) "
58. Categories and commutative algebra "

NOTE: Volumes 1 to 38 are out of print. A few copies of volumes 23,28,31,32,33,34, 36,38 are available on request from C.I.M.E.

1972 -	59.	Non-linear mechanics		Ed Cremonese, Firenze
	60.	Finite geometric structures and their applications		"
	61.	Geometric measure theory and minimal surfaces		"
1973 -	62.	Complex analysis		"
	63.	New variational techniques in mathematical physics		"
	64.	Spectral analysis		"
1974 -	65.	Stability problems		"
	66.	Singularities of analytic spaces		"
	67.	Eigenvalues of non linear problems		"
1975 -	68.	Theoretical computer sciences		"
	69.	Model theory and applications		"
	70.	Differential operators and manifolds		"
1976 -	71.	Statistical Mechanics		Ed Liguori, Napoli
	72.	Hyperbolicity		"
	73.	Differential topology		"
1977 -	74.	Materials with memory		"
	75.	Pseudodifferential operators with applications		"
	76.	Algebraic surfaces		"
1978 -	77.	Stochastic differential equations		"
	78.	Dynamical systems		Ed Liguori, Napoli and Birhäuser Verlag
1979 -	79.	Recursion theory and computational complexity		Ed Liguori, Napoli
	80.	Mathematics of biology		"
1980 -	81.	Wave propagation		"
	82.	Harmonic analysis and group representations		"
	83.	Matroid theory and its applications		"
1981 -	84.	Kinetic Theories and the Boltzmann Equation	(LNM 1048)	Springer-Verlag
	85.	Algebraic Threefolds	(LNM 947)	"
	86.	Nonlinear Filtering and Stochastic Control	(LNM 972)	"
1982 -	87.	Invariant Theory	(LNM 996)	"
	88.	Thermodynamics and Constitutive Equations	(LN Physics 228)	"
	89.	Fluid Dynamics	(LNM 1047)	"

1983 -	90.	Complete Intersections	(LNM 1092)	Springer-Verlag
	91.	Bifurcation Theory and Applications	(LNM 1057)	"
	92.	Numerical Methods in Fluid Dynamics	(LNM 1127)	"
1984 -	93.	Harmonic Mappings and Minimal Immersions	(LNM 1161)	"
	94.	Schrödinger Operators	(LNM 1159)	"
	95.	Buildings and the Geometry of Diagrams	(LNM 1181)	"
1985 -	96.	Probability and Analysis	(LNM 1206)	"
	97.	Some Problems in Nonlinear Diffusion	(LNM 1224)	"
	98.	Theory of Moduli	(LNM 1337)	"
1986 -	99.	Inverse Problems	(LNM 1225)	"
	100.	Mathematical Economics	(LNM 1330)	"
	101.	Combinatorial Optimization	(LNM 1403)	"
1987 -	102.	Relativistic Fluid Dynamics	(LNM 1385)	"
	103.	Topics in Calculus of Variations	(LNM 1365)	"
1988 -	104.	Logic and Computer Science	(LNM 1429)	"
	105.	Global Geometry and Mathematical Physics	to appear	"
1989 -	106.	Methods of nonconvex analysis	to appear	"
	107.	Microlocal Analysis and Applications	to appear	"
1990 -	108.	Recent Developments in Geometric Topology and Related Topics	to appear	"
	109.	Recent Developments in H_∞ Control Theory	to appear	"
	110.	Continua with microstructures	to appear	"
	111.	Mathematical Modelling of Industrical Processes	to appear	"

LECTURE NOTES IN MATHEMATICS

Edited by A. Dold, B. Eckmann and F. Takens

Some general remarks on the publication of monographs and seminars

In what follows all references to monographs, are applicable also to multiauthorship volumes such as seminar notes.

§1. Lecture Notes aim to report new developments - quickly, informally, and at a high level. Monograph manuscripts should be reasonably self-contained and rounded off. Thus they may, and often will, present not only results of the author but also related work by other people. Furthermore, the manuscripts should provide sufficient motivation, examples and applications. This clearly distinguishes Lecture Notes manuscripts from journal articles which normally are very concise. Articles intended for a journal but too long to be accepted by most journals, usually do not have this "lecture notes" character. For similar reasons it is unusual for Ph.D. theses to be accepted for the Lecture Notes series.

Experience has shown that English language manuscripts achieve a much wider distribution.

§2. Manuscripts or plans for Lecture Notes volumes should be submitted (preferably in duplicate) either to one of the series editors or to Springer- Verlag, Heidelberg. These proposals are then refereed. A final decision concerning publication can only be made on the basis of the complete manuscripts, but a preliminary decision can usually be based on partial information: a fairly detailed outline describing the planned contents of each chapter, and an indication of the estimated length, a bibliography, and one or two sample chapters - or a first draft of the manuscript. The editors will try to make the preliminary decision as definite as they can on the basis of the available information. We generally advise authors not to prepare the final master copy of their manuscript (cf. §4) beforehand.

§3. Final manuscripts should contain at least 100 pages of mathematical text and should include

- a table of contents;
- an informative introduction, perhaps with some historical remarks: it should be accessible to a reader not particularly familiar with the topic treated;
- a subject index: this is almost always genuinely helpful for the reader.

§4. Lecture Notes are printed by photo-offset from the master-copy delivered in camera-ready form by the authors. Springer-Verlag provides technical instructions for the preparation of manuscripts, for typewritten manuscripts special stationery, with the prescribed typing area outlined, is available on request. Careful preparation of the manuscripts will help keep production time short and ensure satisfactory appearance of the finished book. For manuscripts typed or typeset according to our instructions, Springer-Verlag will, if necessary, contribute towards the preparation costs at a fixed rate.

The actual production of a Lecture Notes volume takes 6-8 weeks.

§5. Authors receive a total of 50 free copies of their volume, but no royalties. They are entitled to purchase further copies of their book for their personal use at a discount of 33.3 %, other Springer mathematics books at a discount of 20 % directly from Springer-Verlag.

Commitment to publish is made by letter of intent rather than by signing a formal contract. Springer-Verlag secures the copyright for each volume.

<u>Addresses:</u>

Professor A. Dold, Mathematisches Institut, Universität Heidelberg,
Im Neuenheimer Feld 288, 6900 Heidelberg, Federal Republic of Germany

Professor B. Eckmann, Mathematik, ETH-Zentrum
8092 Zürich, Switzerland

Prof. F. Takens, Mathematisch Instituut, Rijksuniversiteit Groningen,
Postbus 800, 9700 AV Groningen, The Netherlands

Springer-Verlag, Mathematics Editorial, Tiergartenstr. 17,
6900 Heidelberg, Federal Republic of Germany, Tel.: (06221) 487-410

Springer-Verlag, Mathematics Editorial, 175 Fifth Avenue,
New York, New York 10010, USA, Tel.: (212) 460-1596

LECTURE NOTES

ESSENTIALS FOR THE PREPARATION OF CAMERA-READY MANUSCRIPTS

Springer
Springer-Verlag
Berlin Heidelberg New York
London Paris Tokyo Hong Kong

The preparation of manuscripts which are to be reproduced by photo-offset require special care. Manuscripts which are submitted in technically unsuitable form will be returned to the author for retyping. There is normally no possibility of carrying out further corrections after a manuscript is given to production. Hence it is crucial that the following instructions be adhered to closely. If in doubt, please send us 1 - 2 sample pages for examination.

General. The characters must be uniformly black both within a single character and down the page. Original manuscripts are required: photocopies are acceptable only if they are sharp and without smudges.

On request, Springer-Verlag will supply special paper with the text area outlined. The standard TEXT AREA (OUTPUT SIZE if you are using a 14 point font) is 18 x 26.5 cm (7.5 x 11 inches). This will be scale-reduced to 75% in the printing process. If you are using computer typesetting, please see also the following page.

Make sure the TEXT AREA IS COMPLETELY FILLED. Set the margins so that they precisely match the outline and type right from the top to the bottom line. (Note that the page number will lie outside this area). Lines of text should not end more than three spaces inside or outside the right margin (see example on page 4).

Type on one side of the paper only.

Spacing and Headings (Monographs). Use ONE-AND-A-HALF line spacing in the text. Please leave sufficient space for the title to stand out clearly and do NOT use a new page for the beginning of subdivisons of chapters. Leave THREE LINES blank above and TWO below headings of such subdivisions.

Spacing and Headings (Proceedings). Use ONE-AND-A-HALF line spacing in the text. Do not use a new page for the beginning of subdivisons of a single paper. Leave THREE LINES blank above and TWO below headings of such subdivisions. Make sure headings of equal importance are in the same form.

The first page of each contribution should be prepared in the same way. The title should stand out clearly. We therefore recommend that the editor prepare a sample page and pass it on to the authors together with these instructions. Please take the following as an example. Begin heading 2 cm below upper edge of text area.

MATHEMATICAL STRUCTURE IN QUANTUM FIELD THEORY

John E. Robert
Mathematisches Institut, Universität Heidelberg
Im Neuenheimer Feld 288, D-6900 Heidelberg

Please leave THREE LINES blank below heading and address of the author, then continue with the actual text on the same page.

Footnotes. These should preferable be avoided. If necessary, type them in SINGLE LINE SPACING to finish exactly on the outline, and separate them from the preceding main text by a line.

Symbols. Anything which cannot be typed may be entered by hand in BLACK AND ONLY BLACK ink. (A fine-tipped rapidograph is suitable for this purpose; a good black ball-point will do, but a pencil will not). Do not draw straight lines by hand without a ruler (not even in fractions).

Literature References. These should be placed at the end of each paper or chapter, or at the end of the work, as desired. Type them with single line spacing and start each reference on a new line. Follow "Zentralblatt für Mathematik"/"Mathematical Reviews" for abbreviated titles of mathematical journals and "Bibliographic Guide for Editors and Authors (BGEA)" for chemical, biological, and physics journals. Please ensure that all references are COMPLETE and ACCURATE.

IMPORTANT

Pagination. For typescript, number pages in the upper right-hand corner in LIGHT BLUE OR GREEN PENCIL ONLY. The printers will insert the final page numbers. For computer type, you may insert page numbers (1 cm above outer edge of text area).

It is safer to number pages AFTER the text has been typed and corrected. Page 1 (Arabic) should be THE FIRST PAGE OF THE ACTUAL TEXT. The Roman pagination (table of contents, preface, abstract, acknowledgements, brief introductions, etc.) will be done by Springer-Verlag.

If including running heads, these should be aligned with the inside edge of the text area while the page number is aligned with the outside edge noting that right-hand pages are odd-numbered. Running heads and page numbers appear on the same line. Normally, the running head on the left-hand page is the chapter heading and that on the right-hand page is the section heading. Running heads should not be included in proceedings contributions unless this is being done consistently by all authors.

Corrections. When corrections have to be made, cut the new text to fit and paste it over the old. White correction fluid may also be used.

Never make corrections or insertions in the text by hand.

If the typescript has to be marked for any reason, e.g. for provisional page numbers or to mark corrections for the typist, this can be done VERY FAINTLY with BLUE or GREEN PENCIL but NO OTHER COLOR: these colors do not appear after reproduction.

COMPUTER-TYPESETTING. Further, to the above instructions, please note with respect to your printout that

- the characters should be sharp and sufficiently black;
- it is not strictly necessary to use Springer's special typing paper. Any white paper of reasonable quality is acceptable.

If you are using a significantly different font size, you should modify the output size correspondingly, keeping length to breadth ratio 1 : 0.68, so that scaling down to 10 point font size, yields a text area of 13.5 x 20 cm (5 3/8 x 8 in), e.g.

Differential equations.: use output size 13.5 x 20 cm.

Differential equations.: use output size 16 x 23.5 cm.

Differential equations.: use output size 18 x 26.5 cm.

Interline spacing: 5.5 mm base-to-base for 14 point characters (standard format of 18 x 26.5 cm).
If in any doubt, please send us 1 - 2 sample pages for examination. We will be glad to give advice.

Vol. 1259: F. Cano Torres, Desingularization Strategies for Three-Dimensional Vector Fields. IX, 189 pages. 1987.

Vol. 1260: N.H. Pavel, Nonlinear Evolution Operators and Semi-groups. VI, 285 pages. 1987.

Vol. 1261: H. Abels, Finite Presentability of S-Arithmetic Groups. Compact Presentability of Solvable Groups. VI, 178 pages. 1987.

Vol. 1262: E. Hlawka (Hrsg.), Zahlentheoretische Analysis II. Seminar, 1984–86. V, 158 Seiten. 1987.

Vol. 1263: V.L. Hansen (Ed.), Differential Geometry. Proceedings, 1985. XI, 288 pages. 1987.

Vol. 1264: Wu Wen-tsün, Rational Homotopy Type. VIII, 219 pages. 1987.

Vol. 1265: W. Van Assche, Asymptotics for Orthogonal Polynomials. VI, 201 pages. 1987.

Vol. 1266: F. Ghione, C. Peskine, E. Sernesi (Eds.), Space Curves. Proceedings, 1985. VI, 272 pages. 1987.

Vol. 1267: J. Lindenstrauss, V.D. Milman (Eds.), Geometrical Aspects of Functional Analysis. Seminar. VII, 212 pages. 1987.

Vol. 1268: S.G. Krantz (Ed.), Complex Analysis. Seminar, 1986. VII, 195 pages. 1987.

Vol. 1269: M. Shiota, Nash Manifolds. VI, 223 pages. 1987.

Vol. 1270: C. Carasso, P.-A. Raviart, D. Serre (Eds.), Nonlinear Hyperbolic Problems. Proceedings, 1986. XV, 341 pages. 1987.

Vol. 1271: A.M. Cohen, W.H. Hesselink, W.L.J. van der Kallen, J.R. Strooker (Eds.), Algebraic Groups Utrecht 1986. Proceedings. XII, 284 pages. 1987.

Vol. 1272: M.S. Livšic, L.L. Waksman, Commuting Nonselfadjoint Operators in Hilbert Space. III, 115 pages. 1987.

Vol. 1273: G.-M. Greuel, G. Trautmann (Eds.), Singularities, Representation of Algebras, and Vector Bundles. Proceedings, 1985. XIV, 383 pages. 1987.

Vol. 1274: N. C. Phillips, Equivariant K-Theory and Freeness of Group Actions on C*-Algebras. VIII, 371 pages. 1987.

Vol. 1275: C.A. Berenstein (Ed.), Complex Analysis I. Proceedings, 1985–86. XV, 331 pages. 1987.

Vol. 1276: C.A. Berenstein (Ed.), Complex Analysis II. Proceedings, 1985–86. IX, 320 pages. 1987.

Vol. 1277: C.A. Berenstein (Ed.), Complex Analysis III. Proceedings, 1985–86. X, 350 pages. 1987.

Vol. 1278: S.S. Koh (Ed.), Invariant Theory. Proceedings, 1985. V, 102 pages. 1987.

Vol. 1279: D. Ieşan, Saint-Venant's Problem. VIII, 162 Seiten. 1987.

Vol. 1280: E. Neher, Jordan Triple Systems by the Grid Approach. XII, 193 pages. 1987.

Vol. 1281: O.H. Kegel, F. Menegazzo, G. Zacher (Eds.), Group Theory. Proceedings, 1986. VII, 179 pages. 1987.

Vol. 1282: D.E. Handelman, Positive Polynomials, Convex Integral Polytopes, and a Random Walk Problem. XI, 136 pages. 1987.

Vol. 1283: S. Mardešić, J. Segal (Eds.), Geometric Topology and Shape Theory. Proceedings, 1986. V, 261 pages. 1987.

Vol. 1284: B.H. Matzat, Konstruktive Galoistheorie. X, 286 pages. 1987.

Vol. 1285: I.W. Knowles, Y. Saitō (Eds.), Differential Equations and Mathematical Physics. Proceedings, 1986. XVI, 499 pages. 1987.

Vol. 1286: H.R. Miller, D.C. Ravenel (Eds.), Algebraic Topology. Proceedings, 1986. VII, 341 pages. 1987.

Vol. 1287: E.B. Saff (Ed.), Approximation Theory, Tampa. Proceedings, 1985–1986. V, 228 pages. 1987.

Vol. 1288: Yu. L. Rodin, Generalized Analytic Functions on Riemann Surfaces. V, 128 pages, 1987.

Vol. 1289: Yu. I. Manin (Ed.), K-Theory, Arithmetic and Geometry. Seminar, 1984–1986. V, 399 pages. 1987.

Vol. 1290: G. Wüstholz (Ed.), Diophantine Approximation and Transcendence Theory. Seminar, 1985. V, 243 pages. 1987.

Vol. 1291: C. Mœglin, M.-F. Vignéras, J.-L. Waldspurger, Correspondances de Howe sur un Corps p-adique. VII, 163 pages. 1987

Vol. 1292: J.T. Baldwin (Ed.), Classification Theory. Proceedings, 1985. VI, 500 pages. 1987.

Vol. 1293: W. Ebeling, The Monodromy Groups of Isolated Singularities of Complete Intersections. XIV, 153 pages. 1987.

Vol. 1294: M. Queffélec, Substitution Dynamical Systems – Spectral Analysis. XIII, 240 pages. 1987.

Vol. 1295: P. Lelong, P. Dolbeault, H. Skoda (Réd.), Séminaire d'Analyse P. Lelong – P. Dolbeault – H. Skoda. Seminar, 1985/1986. VII, 283 pages. 1987.

Vol. 1296: M.-P. Malliavin (Ed.), Séminaire d'Algèbre Paul Dubreil et Marie-Paule Malliavin. Proceedings, 1986. IV, 324 pages. 1987.

Vol. 1297: Zhu Y.-l., Guo B.-y. (Eds.), Numerical Methods for Partial Differential Equations. Proceedings. XI, 244 pages. 1987.

Vol. 1298: J. Aguadé, R. Kane (Eds.), Algebraic Topology, Barcelona 1986. Proceedings. X, 255 pages. 1987.

Vol. 1299: S. Watanabe, Yu.V. Prokhorov (Eds.), Probability Theory and Mathematical Statistics. Proceedings, 1986. VIII, 589 pages. 1988.

Vol. 1300: G.B. Seligman, Constructions of Lie Algebras and their Modules. VI, 190 pages. 1988.

Vol. 1301: N. Schappacher, Periods of Hecke Characters. XV, 160 pages. 1988.

Vol. 1302: M. Cwikel, J. Peetre, Y. Sagher, H. Wallin (Eds.), Function Spaces and Applications. Proceedings, 1986. VI, 445 pages. 1988.

Vol. 1303: L. Accardi, W. von Waldenfels (Eds.), Quantum Probability and Applications III. Proceedings, 1987. VI, 373 pages. 1988.

Vol. 1304: F.Q. Gouvêa, Arithmetic of p-adic Modular Forms. VIII, 121 pages. 1988.

Vol. 1305: D.S. Lubinsky, E.B. Saff, Strong Asymptotics for Extremal Polynomials Associated with Weights on ℝ. VII, 153 pages. 1988.

Vol. 1306: S.S. Chern (Ed.), Partial Differential Equations. Proceedings, 1986. VI, 294 pages. 1988.

Vol. 1307: T. Murai, A Real Variable Method for the Cauchy Transform, and Analytic Capacity. VIII, 133 pages. 1988.

Vol. 1308: P. Imkeller, Two-Parameter Martingales and Their Quadratic Variation. IV, 177 pages. 1988.

Vol. 1309: B. Fiedler, Global Bifurcation of Periodic Solutions with Symmetry. VIII, 144 pages. 1988.

Vol. 1310: O.A. Laudal, G. Pfister, Local Moduli and Singularities. V, 117 pages. 1988.

Vol. 1311: A. Holme, R. Speiser (Eds.), Algebraic Geometry, Sundance 1986. Proceedings. VI, 320 pages. 1988.

Vol. 1312: N.A. Shirokov, Analytic Functions Smooth up to the Boundary. III, 213 pages. 1988.

Vol. 1313: F. Colonius, Optimal Periodic Control. VI, 177 pages. 1988.

Vol. 1314: A. Futaki, Kähler-Einstein Metrics and Integral Invariants. IV, 140 pages. 1988.

Vol. 1315: R.A. McCoy, I. Ntantu, Topological Properties of Spaces of Continuous Functions. IV, 124 pages. 1988.

Vol. 1316: H. Korezlioglu, A.S. Ustunel (Eds.), Stochastic Analysis and Related Topics. Proceedings, 1986. V, 371 pages. 1988.

Vol. 1317: J. Lindenstrauss, V.D. Milman (Eds.), Geometric Aspects of Functional Analysis. Seminar, 1986–87. VII, 289 pages. 1988.

Vol. 1318: Y. Felix (Ed.), Algebraic Topology – Rational Homotopy. Proceedings, 1986. VIII, 245 pages. 1988

Vol. 1319: M. Vuorinen, Conformal Geometry and Quasiregular Mappings. XIX, 209 pages. 1988.

Vol. 1320: H. Jürgensen, G. Lallement, H.J. Weinert (Eds.), Semigroups, Theory and Applications. Proceedings, 1986. X, 416 pages. 1988.

Vol. 1321: J. Azéma, P.A. Meyer, M. Yor (Eds.), Séminaire de Probabilités XXII. Proceedings. IV, 600 pages. 1988.

Vol. 1322: M. Métivier, S. Watanabe (Eds.), Stochastic Analysis. Proceedings, 1987. VII, 197 pages. 1988.

Vol. 1323: D.R. Anderson, H.J. Munkholm, Boundedly Controlled Topology. XII, 309 pages. 1988.

Vol. 1324: F. Cardoso, D.G. de Figueiredo, R. Iório, O. Lopes (Eds.), Partial Differential Equations. Proceedings, 1986. VIII, 433 pages. 1988.

Vol. 1325: A. Truman, I.M. Davies (Eds.), Stochastic Mechanics and Stochastic Processes. Proceedings, 1986. V, 220 pages. 1988.

Vol. 1326: P.S. Landweber (Ed.), Elliptic Curves and Modular Forms in Algebraic Topology. Proceedings, 1986. V, 224 pages. 1988.

Vol. 1327: W. Bruns, U. Vetter, Determinantal Rings. VII,236 pages. 1988.

Vol. 1328: J.L. Bueso, P. Jara, B. Torrecillas (Eds.), Ring Theory. Proceedings, 1986. IX, 331 pages. 1988.

Vol. 1329: M. Alfaro, J.S. Dehesa, F.J. Marcellan, J.L. Rubio de Francia, J. Vinuesa (Eds.): Orthogonal Polynomials and their Applications. Proceedings, 1986. XV, 334 pages. 1988.

Vol. 1330: A. Ambrosetti, F. Gori, R. Lucchetti (Eds.), Mathematical Economics. Montecatini Terme 1986. Seminar. VII, 137 pages. 1988.

Vol. 1331: R. Bamón, R. Labarca, J. Palis Jr. (Eds.), Dynamical Systems, Valparaiso 1986. Proceedings. VI, 250 pages. 1988.

Vol. 1332: E. Odell, H. Rosenthal (Eds.), Functional Analysis. Proceedings, 1986–87. V, 202 pages. 1988.

Vol. 1333: A.S. Kechris, D.A. Martin, J.R. Steel (Eds.), Cabal Seminar 81–85. Proceedings, 1981–85. V, 224 pages. 1988.

Vol. 1334: Yu.G. Borisovich, Yu. E. Gliklikh (Eds.), Global Analysis – Studies and Applications III. V, 331 pages. 1988.

Vol. 1335: F. Guillén, V. Navarro Aznar, P. Pascual-Gainza, F. Puerta, Hyperrésolutions cubiques et descente cohomologique. XII, 192 pages. 1988.

Vol. 1336: B. Helffer, Semi-Classical Analysis for the Schrödinger Operator and Applications. V, 107 pages. 1988.

Vol. 1337: E. Sernesi (Ed.), Theory of Moduli. Seminar, 1985. VIII, 232 pages. 1988.

Vol. 1338: A.B. Mingarelli, S.G. Halvorsen, Non-Oscillation Domains of Differential Equations with Two Parameters. XI, 109 pages. 1988.

Vol. 1339: T. Sunada (Ed.), Geometry and Analysis of Manifolds. Procedings, 1987. IX, 277 pages. 1988.

Vol. 1340: S. Hildebrandt, D.S. Kinderlehrer, M. Miranda (Eds.), Calculus of Variations and Partial Differential Equations. Proceedings, 1986. IX, 301 pages. 1988.

Vol. 1341: M. Dauge, Elliptic Boundary Value Problems on Corner Domains. VIII, 259 pages. 1988.

Vol. 1342: J.C. Alexander (Ed.), Dynamical Systems. Proceedings, 1986–87. VIII, 726 pages. 1988.

Vol. 1343: H. Ulrich, Fixed Point Theory of Parametrized Equivariant Maps. VII, 147 pages. 1988.

Vol. 1344: J. Král, J. Lukeš, J. Netuka, J. Veselý (Eds.), Potential Theory – Surveys and Problems. Proceedings, 1987. VIII, 271 pages. 1988.

Vol. 1345: X. Gomez-Mont, J. Seade, A. Verjovski (Eds.), Holomorphic Dynamics. Proceedings, 1986. VII, 321 pages. 1988.

Vol. 1346: O. Ya. Viro (Ed.), Topology and Geometry – Rohlin Seminar. XI, 581 pages. 1988.

Vol. 1347: C. Preston, Iterates of Piecewise Monotone Mappings on an Interval. V, 166 pages. 1988.

Vol. 1348: F. Borceux (Ed.), Categorical Algebra and its Applications. Proceedings, 1987. VIII, 375 pages. 1988.

Vol. 1349: E. Novak, Deterministic and Stochastic Error Bounds in Numerical Analysis. V, 113 pages. 1988.

Vol. 1350: U. Koschorke (Ed.), Differential Topology. Proceedings, 1987. VI, 269 pages. 1988.

Vol. 1351: I. Laine, S. Rickman, T. Sorvali, (Eds.), Complex Analysis, Joensuu 1987. Proceedings. XV, 378 pages. 1988.

Vol. 1352: L.L. Avramov, K.B. Tchakerian (Eds.), Algebra – Some Current Trends. Proceedings, 1986. IX, 240 Seiten. 1988.

Vol. 1353: R.S. Palais, Ch.-l. Terng, Critical Point Theory and Submanifold Geometry. X, 272 pages. 1988.

Vol. 1354: A. Gómez, F. Guerra, M.A. Jiménez, G. López (Eds.), Approximation and Optimization. Proceedings, 1987. VI, 280 pages. 1988.

Vol. 1355: J. Bokowski, B. Sturmfels, Computational Synthetic Geometry. V, 168 pages. 1989.

Vol. 1356: H. Volkmer, Multiparameter Eigenvalue Problems and Expansion Theorems. VI, 157 pages. 1988.

Vol. 1357: S. Hildebrandt, R. Leis (Eds.), Partial Differential Equations and Calculus of Variations. VI, 423 pages. 1988.

Vol. 1358: D. Mumford, The Red Book of Varieties and Schemes. V, 309 pages. 1988.

Vol. 1359: P. Eymard, J.-P. Pier (Eds.), Harmonic Analysis. Proceedings, 1987. VIII, 287 pages. 1988.

Vol. 1360: G. Anderson, C. Greengard (Eds.), Vortex Methods. Proceedings, 1987. V, 141 pages. 1988.

Vol. 1361: T. tom Dieck (Ed.), Algebraic Topology and Transformation Groups. Proceedings, 1987. VI, 298 pages. 1988.

Vol. 1362: P. Diaconis, D. Elworthy, H. Föllmer, E. Nelson, G.C. Papanicolaou, S.R.S. Varadhan. École d'Été de Probabilités de Saint-Flour XV–XVII, 1985–87. Editor: P.L. Hennequin. V, 459 pages. 1988.

Vol. 1363: P.G. Casazza, T.J. Shura. Tsirelson's Space. VIII, 204 pages. 1988.

Vol. 1364: R.R. Phelps, Convex Functions, Monotone Operators and Differentiability. IX, 115 pages. 1989.

Vol. 1365: M. Giaquinta (Ed.), Topics in Calculus of Variations. Seminar, 1987. X, 196 pages. 1989.

Vol. 1366: N. Levitt, Grassmannians and Gauss Maps in PL-Topology. V, 203 pages. 1989.

Vol. 1367: M. Knebusch, Weakly Semialgebraic Spaces. XX, 376 pages. 1989.

Vol. 1368: R. Hübl, Traces of Differential Forms and Hochschild Homology. III, 111 pages. 1989.

Vol. 1369: B. Jiang, Ch.-K. Peng, Z. Hou (Eds.), Differential Geometry and Topology. Proceedings, 1986–87. VI, 366 pages. 1989.

Vol. 1370: G. Carlsson, R.L. Cohen, H.R. Miller, D.C. Ravenel (Eds.), Algebraic Topology. Proceedings, 1986. IX, 456 pages. 1989.

Vol. 1371: S. Glaz, Commutative Coherent Rings. XI, 347 pages. 1989.

Vol. 1372: J. Azéma, P.A. Meyer, M. Yor (Eds.), Séminaire de Probabilités XXIII. Proceedings. IV, 583 pages. 1989.

Vol. 1373: G. Benkart, J.M. Osborn (Eds.), Lie Algebras, Madison 1987. Proceedings. V, 145 pages. 1989.

Vol. 1374: R.C. Kirby, The Topology of 4-Manifolds. VI, 108 pages. 1989.

Vol. 1375: K. Kawakubo (Ed.), Transformation Groups. Proceedings, 1987. VIII, 394 pages, 1989.

Vol. 1376: J. Lindenstrauss, V.D. Milman (Eds.), Geometric Aspects of Functional Analysis. Seminar (GAFA) 1987–88. VII, 288 pages. 1989.

Vol. 1377: J.F. Pierce, Singularity Theory, Rod Theory, and Symmetry-Breaking Loads. IV, 177 pages. 1989.

Vol. 1378: R.S. Rumely, Capacity Theory on Algebraic Curves. III, 437 pages. 1989.

Vol. 1379: H. Heyer (Ed.), Probability Measures on Groups IX. Proceedings, 1988. VIII, 437 pages. 1989